Design Concepts in Pneumatic Systems

(In the English Units)

Joji Parambath

Design Concepts in Pneumatic Systems
(In the English Units)

Copyright © 2023 Joji Parambath

All rights reserved

ISBN: 9798397873819

https://jojibooks.com

Disclaimer of Liability

The contents of this textbook have been checked for accuracy. We cannot guarantee full agreement since deviations cannot be avoided entirely. Only qualified personnel should be allowed to install and work on pneumatic equipment. Qualified persons are those authorized to commission, ground, and tag circuits, equipment, and systems following established safety practices and standards.

Dedicated to

all my dear friends

Table of Contents

Chapter	Description	Page No
--	Preface	vii
1	Design considerations	1
2	Fundamentals of Pneumatics	4
3	Components of Compressed Air Generation and Storage	12
4	Compressed Air Quality	27
5	Compressed Air Distribution Systems	41
6	Secondary Air Treatment	49
7	Pneumatic Actuators	51
8	Pneumatic Valves	63
9	Energy Efficiency of Pneumatic Systems	66
10	Design of Pneumatic Systems	69
10.1	Design Problem 10.1	73
10.2	Design Problem 10.2	74
10.3	Design Problem 10.3	75
10.4	Design Problem 10.4	76
10.5	Design Problem 10.5	77
10.6	Design Problem 10.6	84
Appendix 1	Air Compressors' Specifications	88
Appendix 2	General Instructions on Pneumatic Actuators	95
Appendix 3	Air Consumption Chart for Industrial-type Tools	96
11	References	98

Preface

In simple terms, a pneumatic system is made up of a compressor station that provides clean and dry compressed air to power pneumatic actuators. To maximize efficiency, the compressor should run at full load. Additionally, the cost of compressed air increases with higher levels of cleaning. The key to designing a cost-effective and efficient pneumatic system is to deliver just enough clean compressed air to meet the demands of consumers.

Designing pneumatic systems requires knowledge of component functions, parameters, and specifications for the power part, control part, and compressed air network. An initial design should be attempted based on requirement specifications, and the design should be refined as necessary. It is critical to synchronize system parameters with the manufacturer's data for optimal design. Further, it is essential to incorporate inbuilt safety into the system.

This book systematically explains the design aspects of pneumatic systems, providing typical examples of designing such systems in the English units for educational and guidance purposes. The knowledge gained may be applied to develop more extensive industrial pneumatic systems. Other textbooks in the fluid power educational series by the same author cover additional fluid power topics, and a list of all books is provided at the end of this book. Visit https://jojibooks.com for further details.

I hope you enjoy reading the book. Please don't hesitate to share your feedback with me at joji.p@hotmail.com to improve the book. Your input is greatly appreciated.

JOJI Parambath
June 2020

Second Edition …

The second edition of the book has been improved with practical insights to ensure it is up-to-date and relevant. Furthermore, a design problem has been included to vividly illustrate the concept of demand-dependent splitting of compressed air delivery across multiple compressors of varying sizes.

JOJI Parambath
June 2023

About the Author

Joji Parambath is an accomplished expert in Pneumatics, Hydraulics, and PLC with an extensive 25-year background in the field. Over the course of his career, he has trained a multitude of professionals from diverse industries, as well as faculty members and engineering students.

Joji is the primary faculty member at Fluidsys Training Centre in Bangalore, India, offering comprehensive training in Pneumatics and Hydraulics. He has authored an impressive 39 books on the subject matter, all designed to convey knowledge on Pneumatics and Hydraulics in a simplistic and easy-to-understand manner.

Joji attributes the creation of his book series to the active engagement and valuable suggestions of his trainees during the training programs. He would like to extend his gratitude towards them.

10th June 2023

Chapter 1 | Design Considerations

A pneumatic system must be designed to meet all the functional requirements of an application with a focus on safety and efficiency. Maximum efficiency can be achieved by operating system components at their peak performance. A good design must aim to keep the overall costs to a minimum.

The system's design should be robust enough to withstand operational hazards and ensure a long life expectancy. It should also facilitate easy maintenance and the efficient removal of contaminants.

Safety is a critical consideration, and the system should be built with interlocks, power-failure locks, and emergency shutdown features to minimize risks.

The design must also consider the required speed of operation, the pressure and temperature ratings, the quality requirements of components, the cost of downtime and component replacement, the sensitivity to contamination, and the environmental conditions.

It is essential to avoid component wear, overload, oversizing, and high cost. Preparing a circuit diagram of the system is also vital.

General Design Principles

Industrial pneumatic systems are designed with correctly-sized components and conductors. Using undersized components and conductors can cause excessive pressure losses resulting from friction. As a consequence, the operating cost increases significantly. On the other hand, the use of oversized components and conductors imposes higher capital and installation costs.

A simple and systematic approach is best for designing a pneumatic system. Figure 1.1 outlines the critical steps involved in the process. These essential steps are: (1) Analyzing and preparing system specifications, (2) Designing the circuit/control system, (3) Selecting and sizing components, (4) Conducting software simulation and analysis, (5) Developing a system prototype, and (6) Evaluating and optimizing system performance.

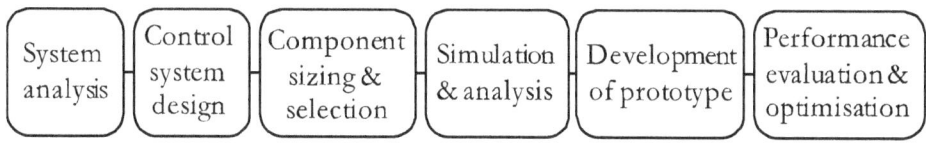

Figure 1.1 | Stages in the design process of a pneumatic system

When designing a pneumatic system, the designer must consider all the unique working conditions of the application. The subsequent sections will outline the necessary steps for designing pneumatic systems.

System Analysis

When designing a pneumatic system, the first step is to define its exact requirements. This involves conducting a detailed system analysis to understand and define the necessary operational specifications and system schematics.

Determining the magnitude of each force (torque) and the types of motions required for the system is crucial. Investigating the types, sizes, and mounting styles of all system actuators is also necessary, as well as determining their air consumption, stroke, duty cycle, and required speed.

Analyzing the type and configuration of air compressors to meet the air requirements of all actuators is a must. The compressor's location, preferably at the load center, must be decided after a study of the surrounding atmosphere's quality.

A study must be conducted on systematically distributing compressed air to all actuators and laying out the distribution system for minimal pressure drop.

Compressed air quality requirements must also be established during the analysis phase, considering the pneumatic system's surrounding environment and climatic conditions.

Various control options must be analyzed depending on the application and control requirements, environmental aspects, and restrictions. It is also necessary to carefully analyze the safety requirements and decide if safety devices, interlocks, power-failure locks, emergency shutdown features, etc., should be incorporated.

Moreover, it is essential to consider the materials for constructing system components and the environment in which the system would operate. Load characteristics, actuator characteristics, sensor characteristics, and the extent of acceptable leakage must also be studied.

Technical parameters of solenoid valves and various electrical control components must be familiarized with, and factors such as temperature, vibration, exposure to outdoor weather, moisture, initial costs, and maintenance costs must be considered.

Manufacturers' technical data must be consulted to configure a solution to the design problem.

Overall, the system's general requirements for robustness, compactness, quality, performance, efficiency, reliability, and safety must be considered. The sequence of operations required in the application must be clearly understood and detailed, and constructing the machine layout of the system may also be necessary.

System Specifications and Circuit Design

Once the requirements of the system are determined, it is important to prepare system and performance specifications.

Following this, appropriate control circuits should be developed using symbols that comply with relevant standards to meet the system specifications.

Finally, alternative circuit solutions should be compared in terms of power transmission efficiency and cost.

Component Selection

It is important to consider key factors to choose the right pneumatic components for a system.

These factors include the application's output requirements, such as force, speed, quality conditions, maintenance needs, and temperature limitations. They determine the main components' type, pressure rating, and size.

For example, the targeted cycle time of work operations in an application is critical when deciding the need for a double-acting cylinder.

The selection of control components will depend on precision, compactness, convenience, safety, and economic needs.

Component Sizing

To ensure a pneumatic system meets specific application requirements, the parameters of its components must be determined. This involves calculating the optimal size or capacity of compressors, valves, actuators, conductors, and accessories based on performance specifications. It is important to note that manufacturers offer graded component sizes, so it is necessary to find suitable near-size components in the market that match the calculated values.

Simulation and Analysis

Using a suitable software package can aid the designer in analyzing how the elements of the system interact and assessing the system's performance through analytical evaluation. A suitable software package can be used to develop and simulate pneumatic or electro-pneumatic circuits. The software allows designers in component- and system-level modeling and assessing steady-state and dynamic responses.

Development of Prototype

Afterward, a system prototype can be created to analyze, assess, and enhance the system's actual performance. The prototype offers a comprehensive view of the system's development, enabling the designer to identify any design flaws and correct them accordingly.

Performance Evaluation

The newly-developed system must meet the required specifications while operating under the specified conditions, particularly in power, torque/force, speed, and efficiency.

Please ensure the system's pressure and flow meet the specified requirements in all operating conditions.

Please verify that all cylinders and motors in the system have sufficient strength to handle their respective loads and accommodate any side loads.

It is important to ensure the system can handle pressure losses, power losses, leakage, and heat generation even under the most challenging operating conditions. If any issues are identified, modifications should be made to optimize the system's performance.

Summary

System Analysis	•Understand the necessary operational specifications and system schematics •Understand the parameters of actuators, compressors, and air distribution system •Understand air quality requirements and control options
System Specification and Circuit Design	•Prepare requirement and performance specifications •Develop control circuits
Component Selection	•Choose the right pneumatic components to satisfy the requirement specifications
Component Sizing	•Calculate the optimal size of compressors, valves, actuators, conductors, and accessories
Simulation and Analysis	•Use a software package to develop and simulate circuits
Development of Prototype	•Create a prototype to analyze, assess, and enhance the system's performance
Performance Evaluation	•Verify that the developed system meets the required specifications

Figure 1.2 | A summary of design considerations

Chapter 2 | Fundamentals of Pneumatics

Pneumatics is the branch of engineering sciences concerned with energy transmission using compressible fluids, like air, etc. Pneumatics is used throughout the industry due to the versatility and simplicity of its application. Many characteristics make pneumatics more appropriate for industrial applications than other power transmission systems.

Gas laws

The gaseous medium in a pneumatic system is sensitive to changes in volume, pressure, and temperature, and the gas laws govern its behavior. Air is a mixture of gases and follows the laws of perfect gas concerning its behavior in volumetric expansion or contraction and absorbing or releasing heat.

Boyle's law

Boyle's law gives the relation between the pressure and volume of a gas. It states that at a constant temperature, the volume of a given mass of gas is inversely proportional to its absolute pressure. Let V_1 (250 in^3) be the volume of a gas at pressure P_1 (14.5 psia) (See Figure 2.1). When the gas is compressed to a volume of V_2 (125 in^3), the pressure will rise to P_2 (29 psia). Mathematically,

$$P_1V_1 = P_2V_2, \text{ where the temperature remains constant}$$

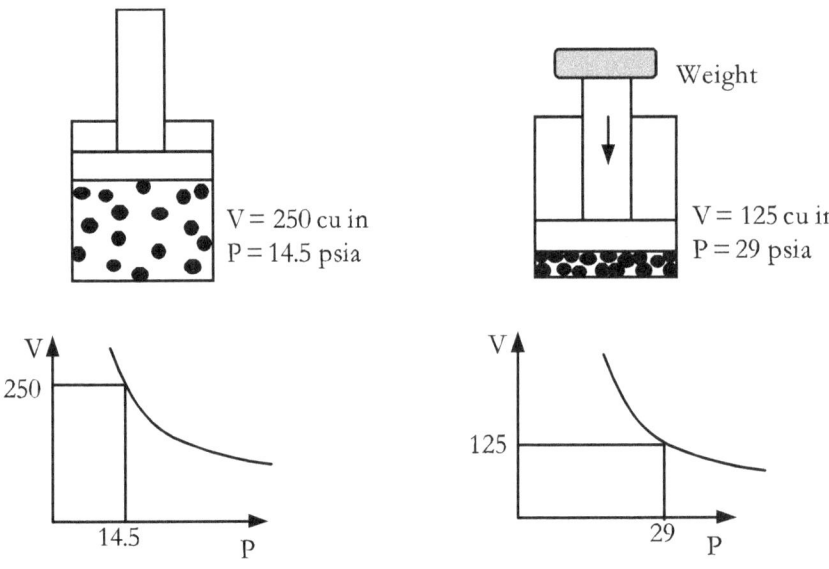

Figure 2.1 | Illustrating Boyle's law

Figure 2.1 illustrates Boyle's law. The development of pressure, as the volume is reduced, is comparatively a slow process. This slow response necessitates using a receiver tank in a pneumatic system to store the compressed gas.

Gay- Lussac's Law

The law states that if the volume of a given mass of gas is held constant, the absolute pressure of the gas varies directly to its absolute temperature.

$$\frac{P1}{T1} = \frac{P2}{T2}; V \text{ remains constant}$$

4

Combined Gas Laws

The general law explains how the variables of absolute pressure, volume, and temperature are related to each other in a fixed mass of gas. The law can be expressed mathematically as:

$$P1 \cdot \frac{V1}{T1} = P2 \cdot \frac{V2}{T2}$$

Air Compression Process

The compression process of air can be thought of as taking place under isothermal, adiabatic, or polytropic conditions.

Isothermal Compression

If the compression of air takes place under a constant temperature condition, the process is said to be isothermal. That means the heat of compression must be removed at the same rate as it is produced. Therefore, the process must be slow enough for the heat to dissipate from the air as it is compressed. The equation governing the isothermal compression can be stated mathematically as follows:

PV is a constant

However, taking out all the heat as it is generated is impossible in practice.

Adiabatic Compression

When a volume of air in a system is compressed or expanded instantly, there is no time to add or dissipate heat into or out of the system, and this type of compression process is said to be adiabatic.

For example, adiabatic compression occurs when air is compressed in a fully-insulated cylinder without any possibility of heat exchange with the surroundings. The same is true with the air expanding through a nozzle very quickly.

The equation governing the adiabatic compression can be stated mathematically as follows:

$P V^n$ is a constant

The value of n for the adiabatic air compression is taken as 1.4.

Polytropic Compression

An isothermal compression process must occur very slowly to keep the air temperature constant. An adiabatic compression process must occur very rapidly without any flow of energy into or out of the system. These compression processes are considered theoretical and presumed to occur under ideal conditions.

In actual practice, air compression occurs between the two limits of compression. A polytropic compression process represents the actual compression process in compressors operating under the normal rate of compression and expansion. For the polytropic compression:

$P V^n$ is a constant

The value of n for a poytropic compression depends on the compression rate and is less than 1.4. Typically, for air, n is taken as 1.3.

Characteristic Curves for the Compression Processes

The characteristic curves for the isothermal, adiabatic, and polytropic compression processes are given in Figure 2.2.

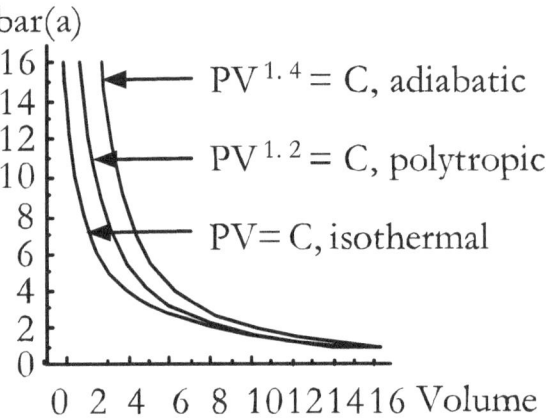

Figure 2.2 Characteristics of compression process for air

Pascal's Law

Pascal's law is central to developing many fluid power devices, such as brakes, presses, and jacks. The law states that pressure at any one point in a static fluid is the same in every direction (Figure 2.3), and pressure exerted on a confined fluid is transmitted equally in all directions, acting with equal force on equal areas.

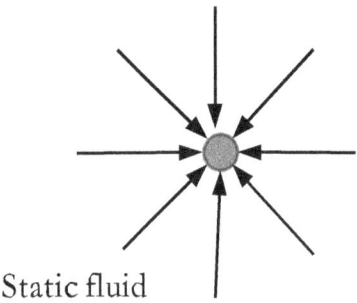

Figure 2.3 | Pascal's law

Pneumatic systems utilize pressurized air to power cylinders and motors. These systems find applications in various industries, such as construction, healthcare, mining, and automotive.

Many kinds of machinery and equipment, such as automated production systems, brake systems for road vehicles and trains, drilling machines, jackhammers, cable jetters, dental drills, and amusement park rides, rely on compressed air as a medium and operate based on Pascal's law.

Air is composed of molecules that are surrounded by a negatively-charged cloud. When air is compressed, the molecules get pushed closer together, which increases their resistance. As a result of this resistance, pressure is generated.

In the next section, we will discuss the concept of pressure and its units.

Pneumatic Pressure

Pressure in pneumatics operates according to Pascal's law. The pressure is the distributed response of force acting through a fluid.

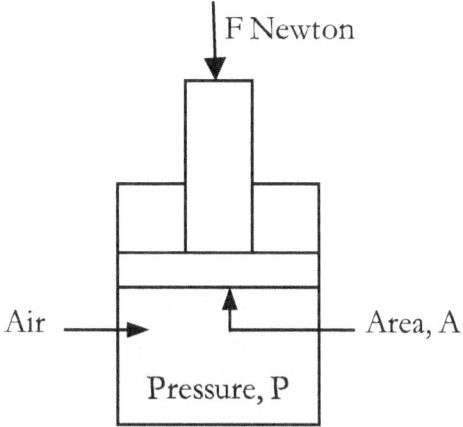

Figure 2.4 | Pressure development in confined air

In Figure 2.4, a definite amount of force (F) is applied to the air enclosed in the chamber through a piston of area A. The enclosed air is compressed, and its pressure (P) rises in direct proportion to the applied force and inverse proportion to the area of the piston. Pressure can, therefore, be defined as the force acting per unit area.

$$P = \frac{F}{A}$$

In the English system, the unit of pressure is pounds per square inch (psi). In the SI system, the unit of pressure is Pascal (Pa), and 1 Pascal is the constant pressure acting on a surface area of 1 square-metre with a perpendicular force of 1 Newton. For industrial pneumatic purposes, Pascal is too small for measurements; hence, more practical units like bar, kilo Pascal, and mega Pascal are used.

1 Pascal	$= 1 \text{ N/m}^2$
1 bar	$= 100000 \text{ Pa} = 10^5 \text{ Pascal}$
1 Mega Pascal (MPa)	$= 10^6 \text{ Pascal} = 10 \text{ bar}$
1 Kilo Pascal (kPa)	$= 10^3 \text{ Pascal}$
1 bar	$= 0.1 \text{ Mpa}$
1 bar	$= 14.5 \text{ Pound per square inch (psi) or [lb/in}^2]$
1 bar	$= 1.02 \text{ kgf/cm}^2$
1 kgf/cm^2	$= 0.981 \text{ bar}$

Pressure scales

Everything on the earth's surface is subjected to a significant pressure head from the weight of the air above. This pressure is the 'atmosphere' (atm) and is approximately equal to 14.7 psi at sea level.

A pressure gauge can measure only the pressure with reference to the local atmosphere. Therefore, the measured pressure value does not include the pressure exerted by the atmosphere. However, at times, we require pressure values with reference to the absolute vacuum.

According to the reference pressure levels, pressures in pneumatic systems can be specified in the following two scales: (1) Gauge scale and (2) Absolute scale.

Gauge Scale

The gauge pressure is indicated by a pressure gauge installed at any location. It is the pressure above the local atmospheric pressure, regardless of the altitude. The gauge pressure, measured in psi, can be stated as psi(g) or simply psi.

Absolute Scale

The absolute pressure scale begins at the point where there is a complete vacuum (zero absolute pressure). The absolute pressure value can be obtained by adding the datum pressure level [for example, 1.013 bar or 14.7 psi at sea level] to the gauge pressure level. The absolute pressure measured in psi should be stated as psi(a).

Remember, absolute pressures are to be used in most of the calculations. Zero gauge pressure indicates the local atmospheric pressure (absolute).

Figure 2.5 illustrates the relationship between the absolute and gauge pressures graphically.

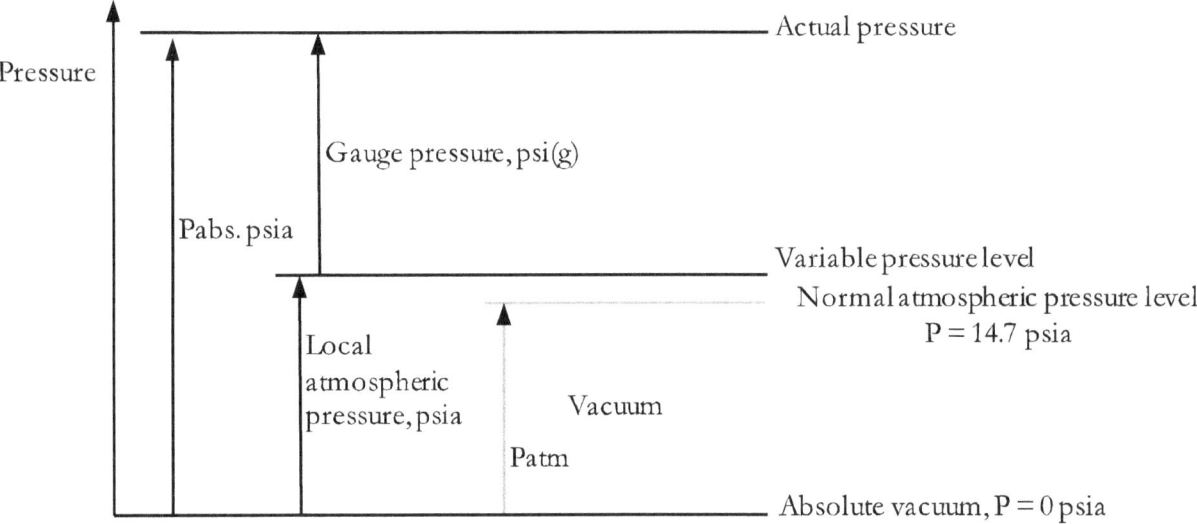

Figure 2.5 | Pressure scales

Economic Pressures in Pneumatics

Pneumatic systems have been developed and progressed comparatively as low-pressure systems, as air compression is a slow process. Pneumatic air-consuming devices such as cylinders and air motors are generally rated for a maximum operating pressure of 115 to 145 psi.

However, practical experience has shown that 90 psi is the ideal pressure for the economic operation of industrial pneumatic systems. This low pressure allows the designer to keep the size of pneumatic components very compact and maintain the cost of the components and piping system to a minimum.

Industrial Pressure Ranges

In most industrial pneumatic systems, the preferred operating pressure range is 90 to 145 psi, as shown in Figure 2.6. Many popular air tools are engineered for pressures between 90 and 100 psi. However, the extended pressure range for industrial pneumatic systems can be up to 230 psi. Control pressures in pneumatics can be as low as 43.5 psi. The corresponding absolute pressure scale is also shown in the Figure.

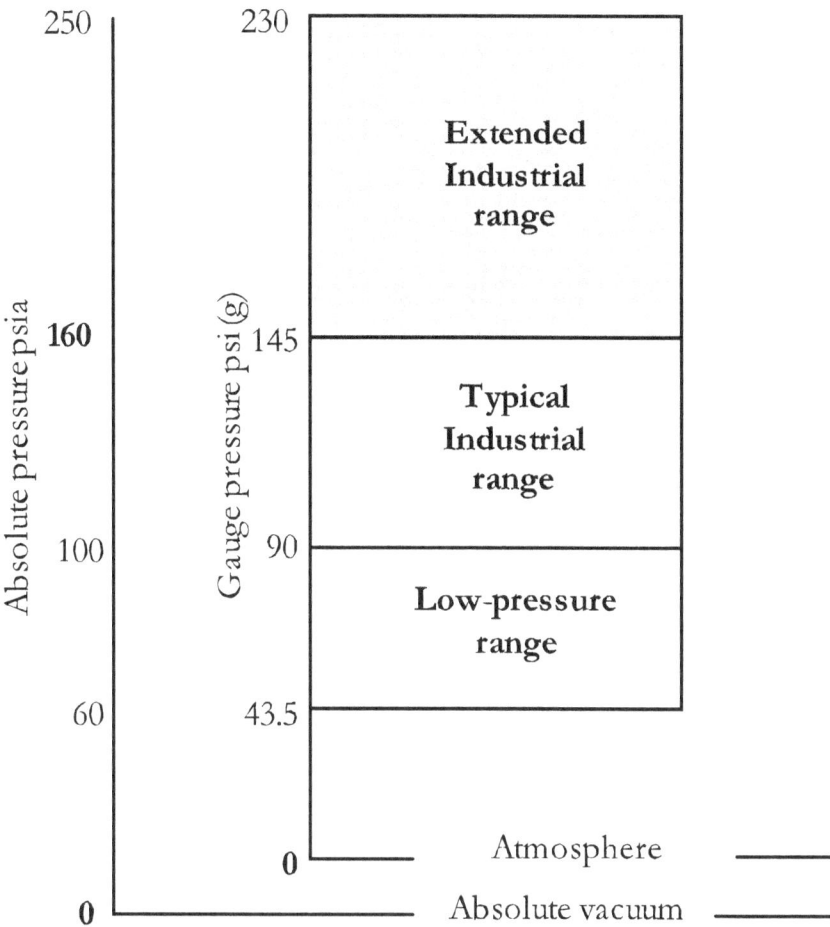

Figure 2.6 | Industrial pressure ranges

Problem 2.1

At atmospheric pressure, 40 cu ft of air is compressed to 6 cu ft. What gauge pressure will be developed if the temperature remains the same?

Solution

Initial volume (V1)	= 40 cu ft
Initial pressure (P1)	= 14.7 psia
Final volume (V2)	= 6 cu ft
Final pressure (P2)	= V1 x P1 / V2
	= 40 x 14.7 / 6
	= 98 psi(a)
	= 83.3 psi(g)

Pneumatic Force

Let us understand how to develop a force to drive a load in the pneumatic system by applying pressure. Figure 2.7 shows the schematic diagram of a pneumatic cylinder with a piston. When the pressure (P) is applied to the area (A) of the piston, it develops a force (F). The amount of force developed equals the applied pressure times the area.

$$\text{That is, } F = P \times A$$

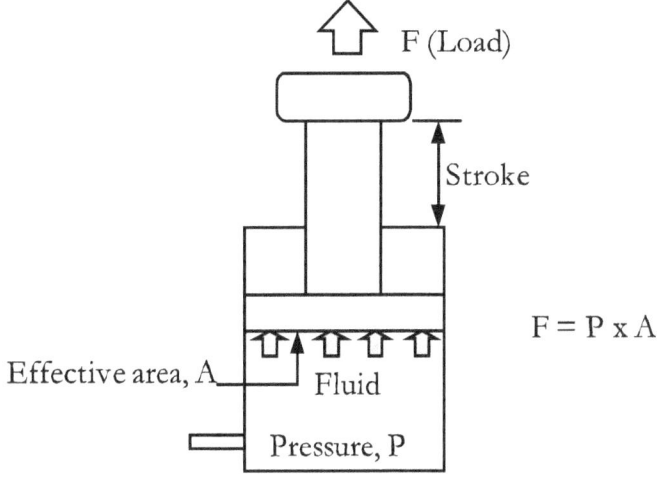

F (Load)

Stroke

$F = P \times A$

Effective area, A

Fluid

Pressure, P

Figure 2.7 | Development of force (F) with the application of pressure (P)

Problem 2.2

Calculate the pressure produced by a force of 1125 pounds acting on the piston with an area of 12.5 in².

Solution

Force	= 1125 lb
Area	= 12.5 in²
Pressure	= F/A
	=1125 / 12.5
	= 90 lb/in² (psi)

Force Multiplication

A pneumatic system can be designed for easy force multiplication. The basic idea of force multiplication is explained with the help of Figure 2.8. It shows an arrangement of two cylinders, A and B, with piston areas A1 (say, 0.06 in²) and A2 (say, 0.6 in²) (A2 > A1), respectively interconnected through a pipe. The enclosed space inside the cylinders and the pipeline is filled with air. When cylinder A is applied with force F1 (2 lb), a pressure P (33.3 psi) is generated in the air medium. The same pressure P acts on the piston of cylinder B, as per Pascal's law. This pressure causes cylinder B to develop a force F2 (20 lb). The governing equations for the forces developed in the cylinders are as follows:

$$F1 = P \times A1$$
$$F2 = P \times A2$$

Therefore,

$$F2 = F1 \times \frac{A2}{A1}$$

Thus a pneumatic system can be designed for force multiplication. The ability of pneumatic systems to realize force multiplication can be considered leverage. However, it may be noted that force multiplication is achieved by sacrificing distance. For example, if cylinder A moves by 4 in, then cylinder B moves by 0.4 in.

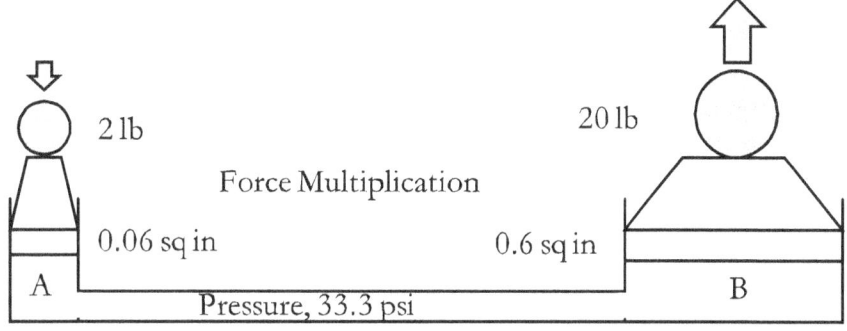

Figure 2.8 Illustration of force multiplication concept

The flow rate of Air

It is the volume of air passing a cross-section per unit of time under the specified pressure, temperature, and relative humidity. It is usually measured in cubic feet per minute (cfm) or liters per minute (lpm).

Problem 2.3

What force is produced by a pneumatic cylinder with an area of 7 in^2 operating at a pressure of 90 psi?

Solution

Area, A	$= 7$ in^2
Pressure, P	$= 90$ psi
Force, F	$= P \times A$
	$= 90 \times 7 = 630$ lb

Problem 2.3

A pneumatic lift arrangement, consisting of a small cylinder of bore diameter 3 in and a large cylinder of bore diameter 8 in, lifts a load of 900 lb. What force is required to be exerted on the piston of the small cylinder to lift the load?

Solution

The bore diameter of the large cylinder	$= 8$ in
The bore diameter of the small cylinder	$= 3$ in
Force to be lifted, F2	$= 900$ lb
Piston area of the small cylinder, A1	$= \prod . 3^2 / 4$
	$= 7$ in^2
Piston area of the large cylinder, A2	$= \prod . 8^2 / 4$
	$= 50$ in^2

Therefore,
Force needs to be exerted on cylinder A1, F1 = F2 . (A1 / A2)

$$= 900 \times (7 / 50)$$
$$= 126 \text{ lb}$$

Chapter 3 | Compressed Air Generation and Storage

The power source in a pneumatic system must be designed to supply a sufficient quantity of compressed air to all the system actuators to get certain work operations. The compressed air medium should also be clean as per the requirements of the associated application.

The primary functions of the power source include the generation and storage of compressed air, regulation of pressure, and removal of heat, solid contaminants, moisture, and oil from the compressed air.

Accordingly, the main components of the power source include a compressor, receiver tank, pressure regulators, aftercooler, filters, and dryer.

A compressor is designed to take in air at atmospheric pressure and deliver it into a closed system at higher pressure to generate the forces needed to perform certain output tasks. The working pressure in the system is regulated using pressure-regulating units.

The compressed air preparation elements, such as aftercoolers, filters, and dryers, remove the undesirable elements, like heat, solid contaminants, moisture, and oil particles, present in the compressed air in various stages of conditioning compressed air.

A distribution system supplies clean and dry compressed air to all the actuators in the system.

In short, the power source must supply clean and dry compressed air at the required pressure and sufficient quantity to all system actuators.

Air Compressors

Compressors are the most common industrial energy supply units. A compressor converts the mechanical power of its prime mover to pneumatic power using a compressible medium. It consists of a moving element enclosed in a housing. It is coupled to a diesel-operated, electric-operated, or gas-operated prime-mover. Electric models are the most popular for indoor applications.

However, it may be noted that the compression process in a compressor is slow. Therefore, a sufficient quantity of compressed air must be stored in a receiver tank.

Pressure Development in a Compressor

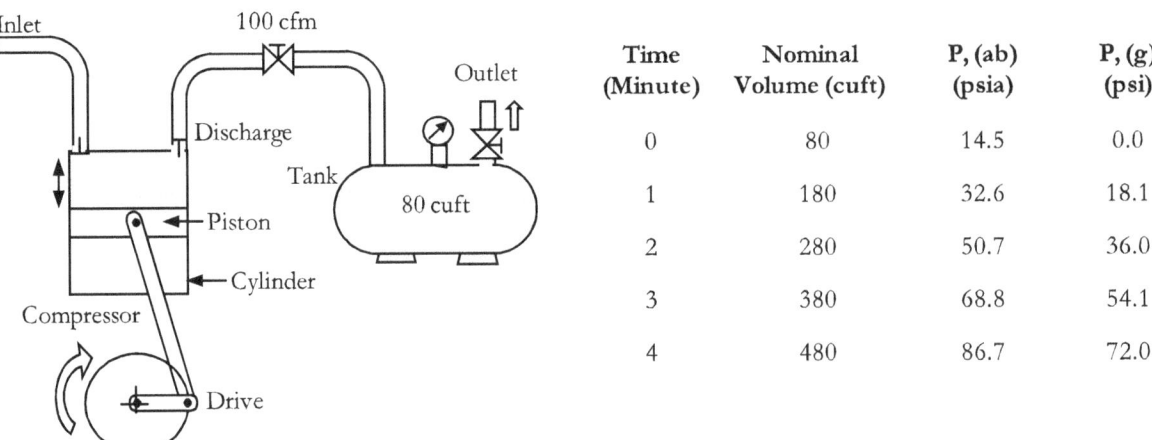

Time (Minute)	Nominal Volume (cuft)	P, (ab) (psia)	P, (g) (psi)
0	80	14.5	0.0
1	180	32.6	18.1
2	280	50.7	36.0
3	380	68.8	54.1
4	480	86.7	72.0

Figure 3.1 | Pressure development in a compressor

Figure 3.1 shows a reciprocating piston compressor connected to a reservoir and a table showing the pressure development with respect to time. It consists of a movable piston enclosed in a cylinder.

Assume that the compressor delivers 3 m³/min of air to the reservoir having a volume of 2 m³.

Using Boyle's law, the pressure rise in the prime-mover-driven compressor can be calculated easily, and the values of absolute and gauge pressures with respect to time are given in the table.

Terms and Definitions, Compressor

A compressor is selected based on the required delivery volume, pressure, and air quality. Remember, parts, parameters, and many other factors concerning a compressor are most important to pneumatic personnel. The essential parameters and factors are its working pressure, operating pressure, air delivery, drive unit type, and pressure regulation and cooling methods. These terms are briefly explained below:

Working Pressure, Compressor

Working pressure is the pressure at the compressor outlet or in the receiver tank. This pressure is usually kept higher than required in the operating position.

Operating Pressure, Compressor

Operating pressure is the pressure that is required at the operating position.

Determining the Operating Pressure and the Working Pressure

The operating pressure can be defined as the pressure needed for the actuators (or consumers) at the point of use. This pressure can be determined by considering the pressure ratings of all actuators. The required operating pressure can be set by using a pressure regulator. If there are three actuators in a pneumatic system, two of which have a pressure rating of 115 psi while the remaining actuator has a pressure rating of 145 psi, then the operating pressure should be set at ≤115 psi.

The working pressure that needs to be developed by the associated compressor can be determined by the pressure drops across various components such as the aftercooler, coarse filters, dryers, pipes, and fine filters between the compressor unit and consumers. Remember that the pressure drop across any component increases as flow increases. Further, a newly installed filter may have a low initial pressure drop that gradually increases as it becomes clogged over time.

Here is an example of how to calculate the working pressure of a system. Let's say the operating pressure is 90 psi, and the pressure drop across each component is as follows: aftercooler (1.45 psi), coarse filter (7.25 psi), dryer (1.45 psi), pipes (2.9 psi), and fine filter (7.25 psi). Using this information, we can calculate the working pressure.

Working pressure = Operating pressure + Pressure drops across all the components

= 90 + 1.45 + 7.25 + 1.45 + 2.9 + 7.25 = 90 + 20.3 = 110.3 psi

Sometimes, only a few consumers require a higher operating pressure (>145 psi), while most require less. To save costs and ensure efficient system operation, it is possible to use one compressor with a low-pressure rating for components operating at lower-pressure levels and another with a high-pressure rating for components operating at higher-pressure levels. This way, all consumers receive the pressure they need without unnecessary expenses.

Conditions of Air

The delivery of compressed air from a compressor, the transmission of compressed air through components and conductors, and the consumption of air by consuming devices in a pneumatic system are expressed in terms of the flow rate. The reference conditions of air must be stated while specifying the flow rate for a uniform representation of the flow rate by the manufacturers of components and end-users of compressed air. The indication of the reference conditions assists the designer to size components accurately.

The reference conditions of air are specified in many ways, as the air can be in different states of pressure, temperature, and humidity. It may be defined as a set of reference conditions of pressure, temperature, and humidity. The significant ways of categorizing air as per the stated reference conditions are: (1) free air, (2) standard air, and (3) normal air.

A designer of pneumatic systems must be familiar with various sets of reference conditions of compressed air for an explicit understanding of many representations of flow rates used in the domain of manufacturers. The definitions of the terms free air, standard air, and normal air are given below.

Free Air

The term free air denotes the air at the atmospheric conditions at the location of a compressor but unaffected by the compressor. This term does not mean air under standard conditions.

Air, under Standard Conditions

The standard set of reference conditions for representing the flow rate of the air delivered, transmitted, or consumed by a pneumatic system is specified in the ISO 1217 standard. This set of conditions is defined as a pressure of 1 bar(a), a temperature of 20° C, and a relative humidity of 0%.

Air, under Normal Conditions

The normal set of reference conditions for representing the flow rate of the air delivered, transmitted, or consumed is defined as a pressure of 1.01325 bar(a), a temperature of 0° C, and a relative humidity (RH) of 0%. The summary of the conditions of the air is given in Table 3.1.

Table 3.1 | Summary of conditions of air

	Pressure	Temperature	Humidity
Free air	Local conditions		
Standard air (ISO 1217)	14.5 psi(a)	68° F	0%
Normal air	14.7 psi(a)	32°F	0%

The flow rate of Air

The flow rate of air, in respect of a compressor, is the volume of air displaced or delivered per unit of time at the rated speed of the driveshaft and under the rated conditions of pressure, temperature, and relative humidity. The flow rate can be measured in terms of: (1) Theoretical flow rate (or displacement) and (2) Effective flow rate (or delivery).

Theoretical flow rate (Displacement Volume)

The theoretical flow rate is the quantity of inlet air a compressor displaces. The theoretical flow rate of a compressor is the product of the volume of air swept in one revolution of its driveshaft and the number of revolutions per unit of time. It is usually expressed as cubic feet per minute (cfm) or liters per minute (lpm).

Also note: 1 cfm 28.32 lpm and 1 lpm = 0.0353 cfm.

Effective flow rate (Delivery Volume)

An effective flow rate is the quantity of air that a compressor delivers at the specified discharge pressure. The discharge pressure is typically specified at 90 psi. The quantity of the delivered compressed air is usually converted back to the actual inlet atmospheric conditions of the compressor at a given site or the standard (or normal) atmospheric conditions to normalize the effective flow rate. Accordingly, the effective delivery volume can be expressed in terms of the actual delivery volume (Free Air Delivery – FAD) or the standard (or normal) delivery volume.

Actual Delivery Volume (Free Air Delivery)

It is the volume of compressed air a compressor delivers at the specified discharge pressure (typically 90 psi) over a period. It is usually stated in terms of the actual prevailing atmospheric inlet conditions. In other words, it is the expanded volume of air it forces into the associated system per unit of time. It is expressed in terms of cfm (fad) or lpm (fad).

Figure 3.2 shows the essential parameters required to calculate free air delivery. To calculate the free air delivery, firstly, the atmospheric pressure (P1), the actual temperature (T1), and the humidity (RH1) at the inlet of the compressor are measured. Next, the maximum working pressure (P2), discharge temperature (T2), and the volume of the compressed air (V2) at the outlet are measured. Pv is the water vapor pressure. Finally, the volume V2 is referred back to the inlet conditions using the ideal gas equation. The value of V1 is the free air delivery of the compressor.

Figure 3.2 | Compressor parameters

$$\text{Free air delivery (FAD, V1)} = \frac{P2 \times V2 \times T1}{[P1 - (Pv \times RH1)] \times T2}$$

Standard (or Normal) delivery volume

It is the volume of compressed air delivered by an air compressor at the specified discharge pressure and is generally stated in terms of the standard (or normal) atmospheric conditions. It is expressed in terms of cfm (std or normal) or lpm (std or normal).

Problem 3.1

How much air under free air conditions can be delivered by a receiver tank of 2 cu ft containing air at a pressure of 90 psi? The temperature inside the tank is 100°F. Neglect Relative Humidity (RH). Assume ambient temperature as 77°F

Solution

P1 = 90 psi = 104.7 psia
V1 = 2 cu ft
T1 = 100°F = 460+100 = 560°R
P2 = 14.7 psia
T2 = 77°F = 460+77 = 537°R

$$V2 = V1 \cdot \frac{P1}{P2} \cdot \frac{T2}{T1}$$

$$V2 = 2 \cdot \frac{104.7}{14.7} \cdot \frac{537}{560} = 13.66 \text{ cu ft (Free air)}$$

Problem 3.2

How much air can be delivered by a receiver tank of 2 cu ft containing air at a pressure of 90 psi under standard air conditions? The temperature inside the tank is 100°F. Neglect relative humidity.

Solution

P_1 = 90 psi = 104.7 psia
V_1 = 2 cu ft
T_1 = 100°F = 460+100 = 560°R
P_2 = 14.5 psia
T_2 = 68°F = 460+68 = 528°R

$$V_2 = 2 \cdot \frac{104.7}{14.5} \cdot \frac{528}{560} = 13.61 \text{ cu ft (Std)}$$

Problem 3.3

How much air under normal air conditions can be delivered by a receiver tank of 2 cu ft containing air at a pressure of 90 psi? The temperature inside the tank is 100°F. Neglect RH.

Solution

P_1 = 90 psi = 104.7 psia
V_1 = 2 cu ft
T_1 = 100°F = 460+100 = 560°R
P_2 = 14.7 psia
T_2 = 32°F = 460+32 = 492°R

$$V_2 = 2 \cdot \frac{104.7}{14.7} \cdot \frac{492}{560} = 12.52 \text{ cu ft (Normal)}$$

Duty Cycle, Compressor

The duty cycle of an air compressor is the amount of time the compressor can run before it needs rest. For example, if the compressor's duty cycle is 50% in one hour, then the compressor is designed to run for 30 minutes in one hour.

Classification of Compressors

Compressors can be classified based on the design of their moving elements, the number of stages of the compression process, and the types of displacements. Compressors can be of lubricated, non-lubricated or oil-less designs. Oil is injected into the compression chamber of a lubricated compressor to lubricate its internal moving elements and bearings. It also takes away most of the compressor's heat due to compression.

Reciprocating Vs Rotary Compressors

Compressors can be classified according to the specific design of the element used to create the flow of air. That is, the reciprocating type or rotary type. Reciprocating piston compressors are very common and provide a wide range of pressures and delivery volumes. However, they are ideally suited for intermittent duty cycles.

Piston compressors are employed where high pressures (60 – 430 psi) and medium delivery volumes (< 600 cfm) are needed. Multi-stage compressors with intercooling between each compression stage are used for higher pressures. Rotary screw or rotary vane types are also used in many industrial applications. They are employed for applications with continuous duty cycles.

Classification: Single-stage Vs Multi-stage Compressors

Figure 3.3(a) shows a single-stage compressor. In the single-stage compressor, an increase in pressure takes place in a single cylinder. Single-stage compressors are generally used for pressures up to 175 psi. Single-stage compressors are adequate for small shops.

In a multi-stage compressor, as shown in Figure 3.3(b), one cylinder's exhaust feeds another's in-stroke to obtain higher outlet pressures. The multi-stage compressor is usually provided with an intercooler to remove the heat of compression. Double-stage compressors can be used for getting pressures up to 435 psi. Multi-stage compressors are more efficient and have a longer service life than single-stage compressors. Two-stage compressors with oversized storage are required for large industrial operations.

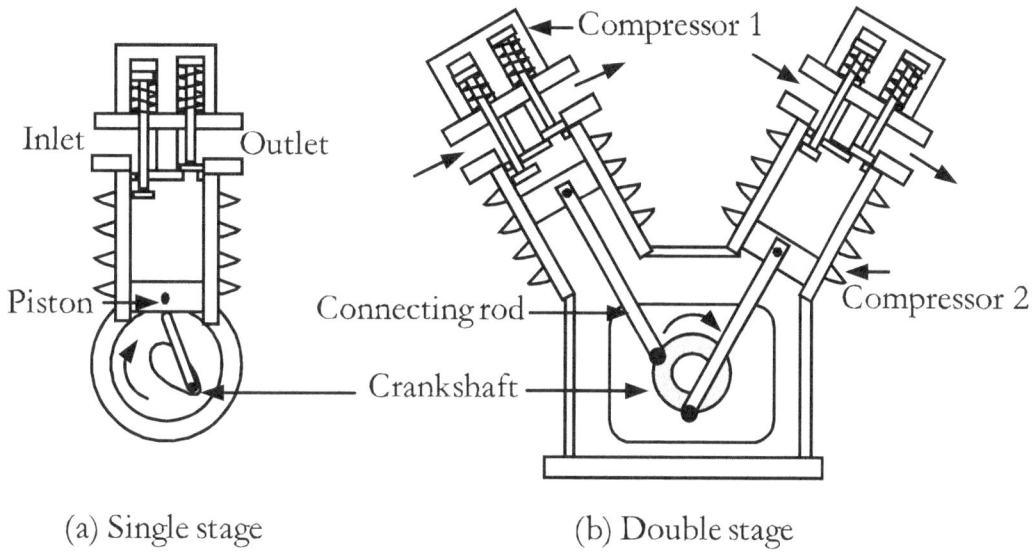

(a) Single stage   (b) Double stage

Figure 3.3 | Single-acting and double-acting compressors

Size Classification, Compressors

Compressor sizes range from a small compressor generating less than 0.6 cfm with little or no preparation equipment to multiple compressor plant installations generating hundreds of cfm.

Small compressors have delivery volumes up to 70 cfm and drive powers up to 20 hp. Medium-sized compressors have delivery volumes ranging from 85 to 600 cfm and drive powers between 20 to 130 hp. On the other hand, compressors exceeding the medium-size limit are classified as large compressors. You can find a summary of these size classifications in Table 3.2.

Table 3.2 | Typical size classification of compressors

Size classification	Delivery volume cfm (FAD)	Drive power hp
Small compressor	<70	<20
Medium-sized	70 – 600	20 – 130
Large compressor	>600	>130

Positive Vs Dynamic Displacement Compressors

A broad classification of compressors, according to the displacement, is shown in Figure 3.4. Compressors are generally classified according to the compressing elements used as: (1) positive displacement devices and (2) dynamic displacement devices.

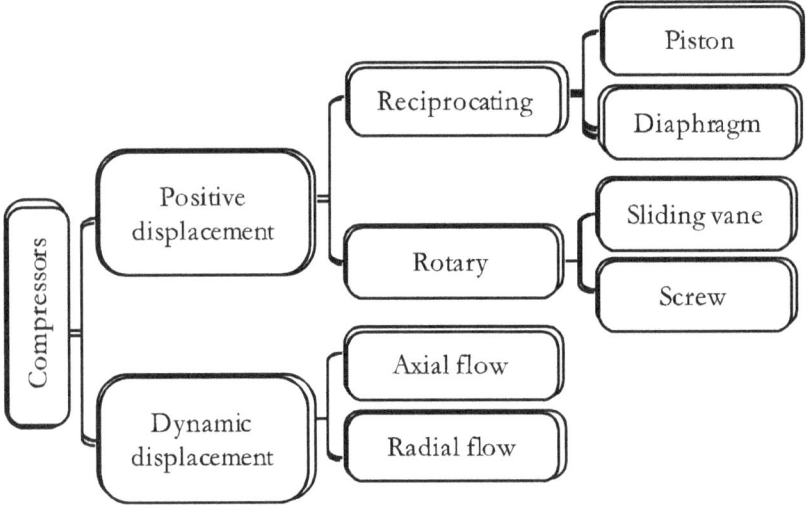

Figure 3.4 | Classification of compressors

In a positive displacement compressor system, the air is confined within an enclosed space where it is compressed by decreasing its volume. In a dynamic displacement compressor, the air is accelerated by the rapidly rotating elements, such as rotor blades, causing some increase in pressure and a significant increase in velocity.

Reciprocating Piston Compressor

Figure 3.5 shows the basic single-cylinder reciprocating compressor. As the piston moves during its inlet stroke, the inlet valve opens and draws air into the cylinder. During the outstroke of the piston, the air is compressed and discharged through the outlet valve. Piston compressors have a relatively complex design with many moving parts.

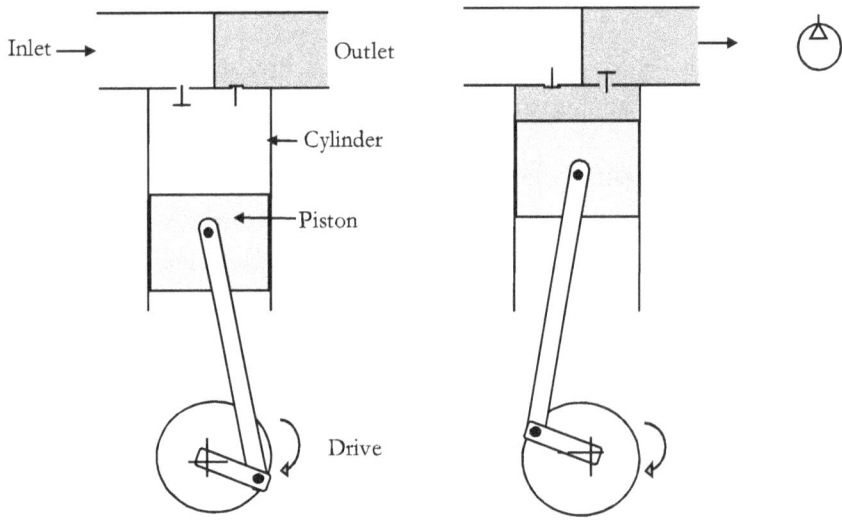

Figure 3.5 | Reciprocating compressor

Diaphragm Compressor

In piston compressors, there is a likelihood of introducing small amounts of lubricating oil from the piston walls into the compressed air. This minimal oil contamination may be unwanted in food, pharmaceutical, and chemical industries and hospital and laboratory applications. Diaphragm compressors may be used as power sources for such applications. Figure 3.6 shows a diaphragm compressor. A flexible diaphragm separates the compressor chamber and the actuating piston. This feature excludes the lubricating oil from the compressed air supply. Diaphragm compressors have limited delivery and pressure levels. They are used most often for light-duty applications.

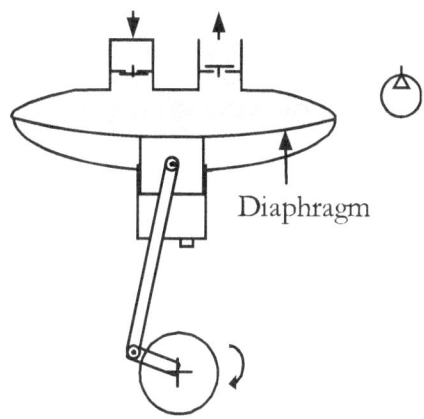

Figure 3.6 | Diaphragm compressor

Screw Compressor

A screw compressor is shown in Figure 3.7. It consists of two helically grooved screws meshing with each other with a little clearance of about 0.002 in. The design of the screws makes it possible to move air from the inlet to the outlet of the compressor. Compression is achieved by pushing the trapped air into a progressively smaller volume as the screws move ahead. Since no surfaces make contact with one another, this type of compressor does not necessitate cooling and is characterized by a low noise level and a small loss of efficiency. They have the benefit of simplicity with fewer moving parts rotating at a constant speed and of a steady delivery of compressed air without pressure fluctuations.

Single-stage screw compressors are generally designed to operate at pressures less than 150 psi and capacities up to 200 cfm. Higher pressures and capacities can be attained by multi-stage compression.

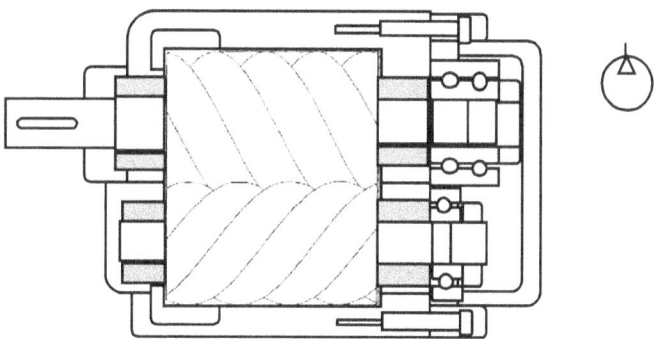

Figure 3.7 | Screw compressor

Rotary screw compressors are widely used in various industries due to their reliability, energy efficiency, and low noise levels. They are particularly useful when a large volume of compressed air and continuous flow are required, such as in automated manufacturing and food packaging plants.

Vane Compressors

It consists of a prime-mover-driven rotor with sliding vanes in close-fitting radial slots, as shown in Figure 3.8. The rotor moves within a larger circular cavity. The centers of the rotor and the cavity are offset by a certain distance, causing an eccentricity. The vane tips bear against the casing and form an adequate seal. Side plates keep the fluid confined to the existing space along the width of the rotor and vanes. Oil is injected into the compression chamber as a lubricant and a seal. It also removes the heat of compression.

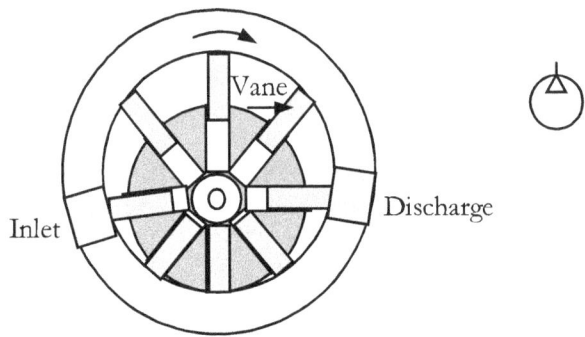

Figure 3.8 | Sliding vane compressor

As the rotor rotates, the space between two successive vanes increases at the suction side. This expanding volume creates a partial vacuum, which draws air into the chambers formed by the vanes. The air is trapped in these chambers. The trapped air is then moved through the compressor by the rotating vanes. Compressed air is squeezed through the discharge port as the space between the two rotating vanes decreases. Sliding vane compressors have power ratings ranging from 10 to 200 hp, flow rate capacities from 40 to 800 cfm, and discharge pressures from 80 to 125 psig.

Drive, Compressor

Electrical motors or IC engines drive compressors. The electric drive can be single-phase or three-phase type. In factories, compressors are driven by 3-phase or single-phase induction motors. Power is transmitted through V-Belt, gear, or direct drive configurations. Traditional V-Belt drives provide great flexibility in coupling the compressor and its prime mover at minimum cost. A gear drive reduces the axial load on the moving element of a compressor, thus extending its operational life. Gear drives typically use less energy than V-belt drives. A direct drive can offer a compact configuration and minimum maintenance. Belts and coupling used in the compressor system must be adequately shielded for safety.

Cooling of Compressors

Cooling fins on smaller compressors permit the heat to be removed by radiation. A large compressor usually has an additional fan to remove the heat by forced air cooling. In the case of a compressor plant with a drive power over 40 hp, the forced air-cooling is inadequate. The compressors are then installed with a water-circulation cooling system.

Constructional Features of Compressors

Compressors are available for light-duty or heavy-duty applications. They are typically constructed with a cast-iron cylinder and hardened steel crankshaft with precision bearings.

Further, they are available as oil-free or lubricated types. A compressor can be of the open-frame type construction or the totally-enclosed type construction. The totally-enclosed type of construction tends to reduce compressor noise.

Next, the drive part may also be provided with an enclosed belt guard.

Storage of Compressed Air

The use of an air receiver is a simple method of power storage. Sufficient storage of compressed air is crucial for properly functioning a compressed air system. It represents the available energy that can be released when needed. A properly-sized receiver tank can provide sufficient storage capacity to meet any peak demands of compressed air and prevent excess start-stop cycling in the on-off pressure regulation of the system compressor. The size of a receiver tank is determined by several factors, including the delivery volume of the compressor, load requirements, the acceptable pressure drop from the compressor station to the point of use, the time required for the compressor to start up, and the duration available for refilling the stored compressed air. The details of air receivers are given in subsequent sections of this chapter.

Typical Specifications, Compressors

The most important specifications for air compressors include the number of stages, compressed air delivery, maximum pressure rating, duty cycle, motor power, and operating voltage.

Some essential parameters of single-stage and multi-stage compressors and their values are given in Table 3.3 and Table 3.4, respectively.

Single-stage (Maximum Pressure 125 psi), Typical Specifications

Table 3.3 | Single-Stage compressors (Max Pressure 125 psi)

Power	Drive speed	Displacement	FAD @90psi	Air Receiver
hp	rpm	cfm	cfm	gallon
3	550	12.7	9	40
5	925	21.4	15.6	60
7.5	690	32.6	25	60
10	920	43.4	33.6	60
10	920	43.4	33.6	110
15	925	66.4	49.1	130
20	925	85	68	120
25	1050	108	87	120

Two-Stage (Maximum Pressure 175 psi), Typical Specifications

Table 3.4 | Two-Stage compressors (Max Pressure 175 psi)

Power	Drive speed	Displacement	FAD @90psi	Air Receiver
hp	rpm	cfm	cfm	gallon
3	925	10.7	8.8	40
3	925	10.7	8.8	60
5	925	17.7	14.5	60
7.5	1050	24.8	20.7	60
10	750	35.4	30.1	60
10	750	35.4	30.1	110
10	750	35.4	30.1	130
15	1150	54.3	44.1	130
20	1150	77.5	62	120
25	1150	105	84	120
30	1150	118	95	120

A rule of thumb suggests that on a steady pumping, a compressor will produce a minimum of 4 scfm flow of air for every hp capacity at 90 psi.

Note: Data for different types of air compressors are given in Appendix 1

Air Receivers, Construction

Using an air receiver (tank) is a simple power storage method. Figure 3.9 shows an air receiver with essential constructional features.

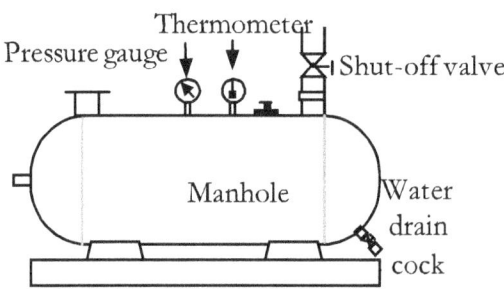

Figure 3.9 | Air receiver tank

Air receivers are available as horizontal models and vertical models. A vertical model is used when the floor space for its installation is limited. Air receivers must be provided with manual or automatic drains for releasing the collected water. They are usually made of mild steel.

The size of a compressed air receiver depends on the delivery volume of the compressor, load requirements, and the allowable pressure deviations in the receiver. Typical tank sizes include: 1.2, 2.3, 2.5, 2.6, 3, 3.2, 4, 4.5, 5, 6, 8, 9, 10, 15, 17, 20, 24, 26, 30, 60, 80, 120, 200, 240, 400, and 660 gallons.

Air receivers must be fabricated as per the relevant standards in one's region. Conformity to the standards ensures that the tank plates have sufficient thickness and are made with proper materials. During the fabrication, proper welding techniques must be applied by experienced operators.

An air receiver is also provided with (1) a safety relief valve to guard against high pressures, (2) a pressure switch to sense the air pressure inside the tank, (3) a high-temperature switch to sense the excess air temperature inside the tank, and (4) pressure gauges for pressure indication.

Compressor Control / Pressure Regulation

Compressors supply clean compressed air in sufficient quantity and at the required pressure to meet the compressed air demand of a system. A compressor with its drive motor works most efficiently under its full-load operation. However, compressed air demand and pressure may vary depending on the system's work operations. When the demand or the load factor fluctuates, the energy efficiency of the compressor system tends to reduce.

As you are aware, working pressure in a receiver is generally set higher than what is required at the operating position. A compressor can be controlled by its working pressure. Usually, when there is a surge in demand, the pressure drops, and vice versa. Sensing the pressure and using an appropriate control system can minimize the fluctuations in pressure. A control system can be designed to increase the compressor output when there is a pressure drop and decrease the compressor output when there is a pressure rise. Compressors use various capacity control systems to always provide the correct amount of compressed air supply to the fluctuating compressed air demand and regulate the pressure fluctuations within safe limits. A popular and efficient method of capacity control of a compressor is the on-off control of its drive motor, which runs the compressor at full load or no load according to the fluctuations in the compressed air demand. Another method to control the compressor delivery is to vary the speed of the compressor drive motor using variable frequency converters.

On-off Control, Compressors

The drive motor of a compressor is switched on (loading) when the working pressure drops to a lower limit P_{min} and the compressor runs at full load. And the drive motor is switched off (off-loading) when the working pressure reaches an upper limit P_{max}. Pressure variations normally lie within the 4 to 14.5 psi range.

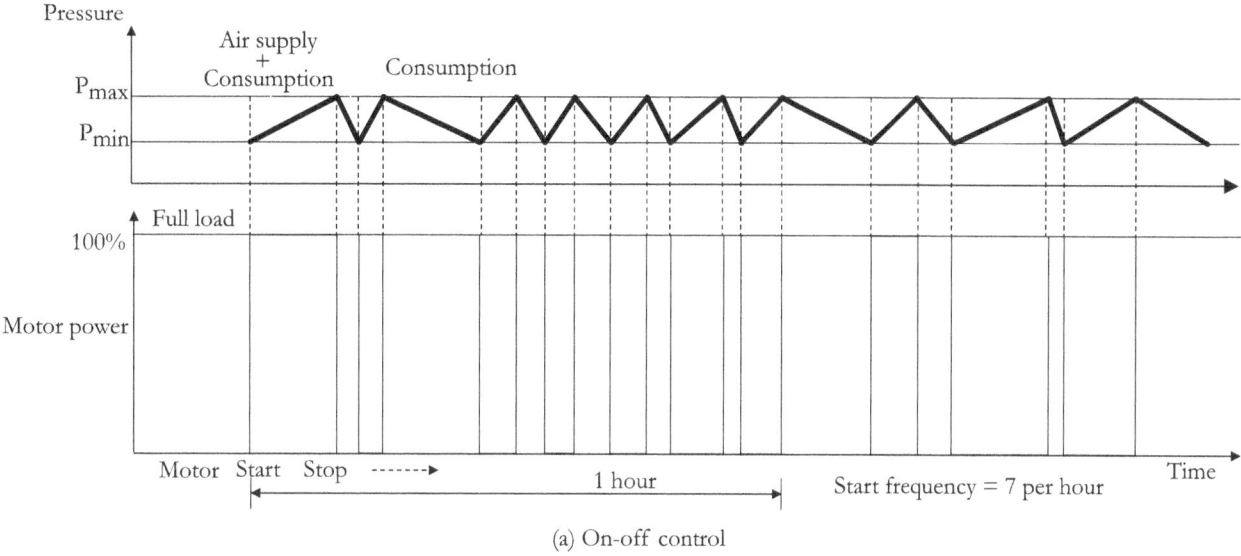

(a) On-off control

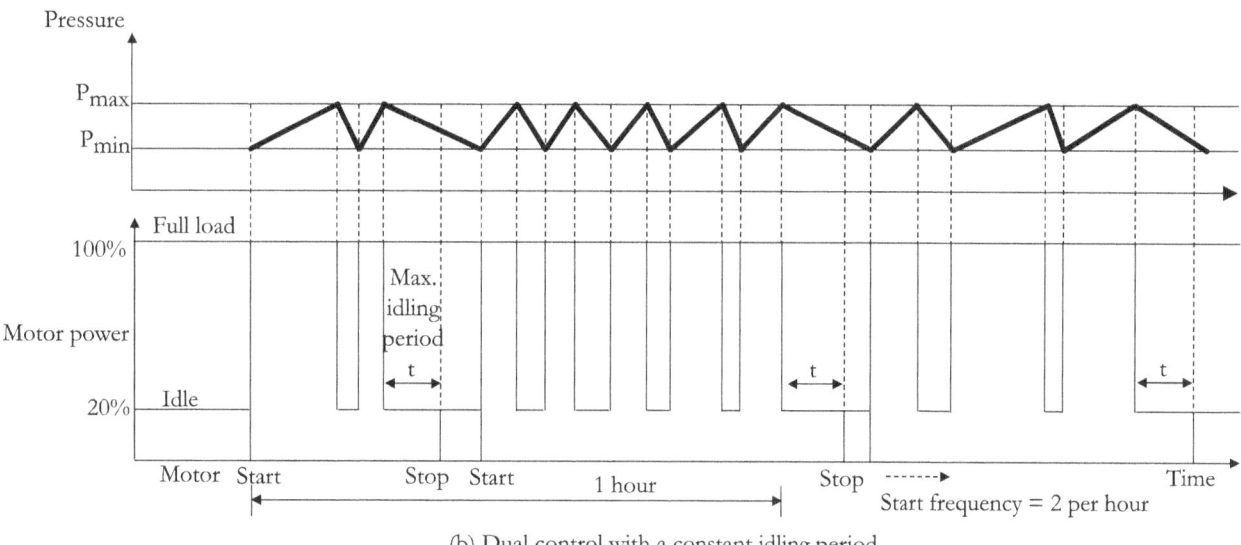

(b) Dual control with a constant idling period

Figure 3.10 | Compressor controls

However, it may be noted that switching on or switching off motors is a harmful operation, especially for large motors, and manufacturers restrict the switching frequency of motors. This can be achieved by running the drive motor for some more time but with no-load compressor output when the maximum pressure P_{max} is reached. The extended running of the drive motor is the idling period of the motor. The power required to run the motor during the idling period must be considered a power loss.

Therefore, various methods like dual control with a constant idling period, automatic optimal operating mode selection for controlling the idling period, temperature-dependent control of the idling period, etc., are devised to control the switching frequency of motors. Figure 3.10 illustrates the concepts of some of these methods.

Safety Relief Valve
A compressor is usually equipped with a safety relief valve to protect against any failure of its pressure regulation system.

Constructional Features, Air Receivers
Air receivers are constructed with a steel tank in vertical or horizontal designs. They typically have a regulator, pressure switch, safety valve, fan-cooled motor, pressure gauge, and manual or automatic drain.

Sizing of Air Receivers
The compressor tank size can be determined based on the type of usage. If the usage is in short quick, concentrated bursts, then a small tank size can be used. If the unit is to sustain long usage periods, a larger tank is required. The size of a receiver tank (V) can be determined using the following formula:

$$\text{Receiver size, V} = \frac{P_a \times t \times (Q_r - Q_c)}{P_{max} - P_{min}}$$

Where,

V	= Size of the receiver tank, ft^3
t	= Time to go from the upper to lower pressure limits, min
Q_r	= Consumption rate of air, scfm
Q_c	= Compressor delivery rate, scfm
P_{max}	= Maximum pressure in the receiver, psia
P_{min}	= Minimum pressure in the receiver, psia
P_a	= Atmospheric pressure, psia

Problem 3.4
Calculate the minimum size of a tank that must supply air to a pneumatic system consuming 1000 scfm for 5 seconds between 110 and 100 psi before the compressor resumes operation.

Solution

Q_r = 1000 scfm | Q_c = 0
t = 5 sec = 5/60 min = 0.0833 min
P_{max} = 110 psi
P_{min} = 100 psi

$$\text{Receiver size, V} = \frac{P_a \times t \times (Q_r - Q_c)}{P_{max} - P_{min}}$$

$$\text{Receiver size, V} = \frac{14.7 \times 0.0833 \times (1000 - 0)}{110 - 100}$$

$$= 122.45 \text{ cu ft}$$

Problem 3.5

Calculate the required size of a tank that must supply air to a pneumatic system consuming 100 scfm for 5 minutes between 115 psi and 100 psi if the compressor is running and delivering air at 70 scfm.

Solution

$$Q_r \quad = 100 \text{ scfm}$$
$$Q_c \quad = 70 \text{ scfm}$$
$$t \quad = 5 \text{ min}$$
$$P_{max} \quad = 115 \text{ psi} \mid P_{min} = 100 \text{ psi}$$

$$\text{Receiver, size, V} = \frac{14.5 \times t \times (Q_r - Q_c)}{P_{max} - P_{min}}$$

$$\text{Receiver, size, V} = \frac{14.5 \times 5 \times (100 - 70)}{115 - 100}$$

$$= 145 \text{ cu ft}$$

Sizing Air Receiver Tank Using Charts

An air receiver tank can also be approximately sized by referring to charts. The sizing based on the airflow capacity is given in Table 3.5, and the sizing based on the compressor power is given in Table 3.6.

Table 3.5 | Receiver tank size Vs airflow capacity

Airflow capacity, cfm	Receiver tank size, gal
10	10
20	20
30	30
40	40
50	50
75	75
100	100
150	150
200	200

Table 3.6 | Receiver tank size Vs compressor power

Compressor power, hp	Receiver tank size, gal
5	20
7.5	30
10	40
15	60
20	80
25	100
30	120
40	160
50	200
60	240
75	300
100	400

1 cu ft = 7.48 gallons

Air Compressor Packaged Units

Most industrial air compressors are supplied as self-contained packages.

An air compressor packaged unit is a fully assembled compact air compressor system with an air compressor, receiver tank, inlet filter, electric motor, belt/gear/direct drive, and automatic microprocessor controllers.

An electronic controller is provided for the intelligent shutdown of the unit and energy-saving operation.

Next, optional equipment includes an aftercooler, particulate filters, an integrated dryer, an automatic moisture drain, a low oil safety control, a magnetic starter, a cooling fan, and a pressure-reducing valve.

The packaged units have low noise enclosures (65 – 70 dBA) and vibration isolators for quiet operation.

Sizing an Air Compressor

- To create an efficient compressed-air system, consider the compressor's size and pressure rating and the arrangements for preparing and distributing the compressed air.

- Sizing an air compressor requires the determination of the logical sequence of steps and the average air consumption rate for a given application.

- Decide the voltage and phase of the electric supply depending on the region where the compressor will be located and the drive power.

- Determine the compressor tank (receiver) size. Most manufacturers offer standard sizes based on the CFM of the compressor (the most popular sizes are 80, 120, and 240 gallons).

Chapter 4 | Compressed Air Quality

Contaminants can enter a pneumatic system through the air taken in by the system compressor. In industrial surroundings, air carries many solid impurities and moisture. Contaminants are harmful to pneumatic systems. Therefore, compressed air discharged from a compressor must be sufficiently prepared, as per a quality standard, to meet the requirements of the associated application.

Solid Contaminants
Solid impurities are industrial dust, including iron, carbon, silicates, fiberglass, soot, and other abrasive materials.

Humidity
Humidity is the moisture present in the atmosphere. Moisture is in the form of water vapor that remains suspended in the given volume of air. It is difficult to remove the moisture from the air. Humidity is usually expressed in terms of either absolute humidity or relative humidity.

Absolute Humidity
Absolute humidity is the amount of moisture in 1000 ft^3 of air.

For example, as shown in Figure 4.1(a), if 0.79 lb of moisture is present in 1000 ft^3 of air at a particular temperature, say at 80°F, then its absolute humidity is 0.79 lb per 1000 ft^3 of air at 80°F. It is always temperature-dependent.

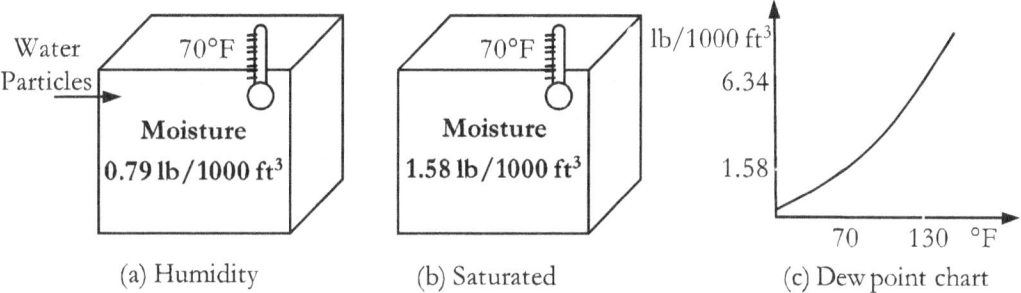

(a) Humidity   (b) Saturated   (c) Dew point chart

Figure 4.1 | Absolute humidity concept

Saturation Quantity
A given volume of air at a specified temperature can contain moisture in the vapor form up to its saturation level.

For example, as shown in the dew point chart of Figure 4.1(b), the following can be observed. 1000 ft^3 of free air can contain a maximum of 1.58 lb of moisture at 68°F. However, at 130°F, it can contain a maximum of 6.34 lb of moisture.

The saturation quantity is a function of temperature and is given by the dew point chart [See Figure 4.1(c)].

Relative Humidity
The relative humidity (RH) of air is the ratio of its absolute humidity to the air saturation quantity at a given temperature. It is usually expressed as a percentage. That is,

For example, if 1000 ft³ of air contains 0.79 lb of moisture at 68°F, then the relative humidity can be calculated as follows:

Saturation quantity at 68°F (From Dew point chart) = 1.58 lb

$$RH = \frac{\text{Absolute humidity}}{\text{Saturation quantity}} \times 100\% = \frac{0.79}{1.58} \times 100 = 50\%$$

It may be noted that 100% RH means the given volume of air is saturated. The relative humidity is dependent on both temperature and pressure.

Decreasing the temperature or increasing pressure will condense excess moisture above saturation.

Oil Particles

Oil is used as a lubricating or working medium in many industrial machines. Therefore, air carries harmful oil particles that can be far more than 10 mg/m³ in industrial surroundings.

Air Quality Classification

ISO 8573-1: 2010 stipulates contaminants and quality classes of compressed air for general use. Air contains solid, water, and oil particles as contaminants. The standard specifies the amount of contamination allowed in each cubic meter of compressed air. A quality (or purity) class number is defined for each contaminant according to the permissible levels of a specific parameter(s). These parameters and their permissible values against each class are given in Table 4.1.

Table 4.1 | Permissible levels of contaminants as per ISO 8573-1 (2010)

| Class | Solid particulates | | | | Water | | Oil |
| | Max. particles per m³ | | | Mass concentration, mg/m³ | Vapor pressure dew point | Liquid water g/m³ | mg/m³ |
	0.1 – 0.5 micron	0.5 – 1 micron	1 – 5 micron				
0	As per a written specification between the user and the supplier (more stringent than class 1)						
1	≤20000	≤400	≤10	-	≤-94°F	-	0.01
2	≤400000	≤6000	≤100	-	≤-40°F	-	0.1
3	-	≤90000	≤1000	-	≤-4°F	-	1
4	-	-	≤10000	-	≤37°F	-	5
5	-	-	≤100000	-	≤45°F	-	-
6	-	-	-	≤5	≤50°F	-	-
7	-	-	-	5-10	-	≤0.5	-
8	-	-	-	-	-	0.5-5	-
9	-	-	-	-	-	5-10	-
x	-	-	-	>10	-	>10	>10

An air quality class is a combination of the three air quality numbers. For example, a quality class 1.2.1 means that in each m³ of compressed air, the particulate count should not exceed 20000 particles in the 0.1-0.5 μ size range, 400 particles in the 0.5-1 μ size range and 10 particles in the 1-5 μ size range. A pressure dewpoint of -40°C or better is required, and no liquid water is allowed. In each cubic meter of compressed air, not more than 0.01mg of oil is allowed. This oil level is the total level for liquid oil, oil aerosol, and oil vapor.

Preparation of Compressed Air

The compressed air a compressor delivers has many harmful contaminants and objectionable conditions. The contaminants and conditions are enumerated below:

- The compressed air is very hot.
- It contains a very high concentration of solid particles
- It contains moisture in the vapor and liquid forms
- It contains oil in the liquid and vapor forms

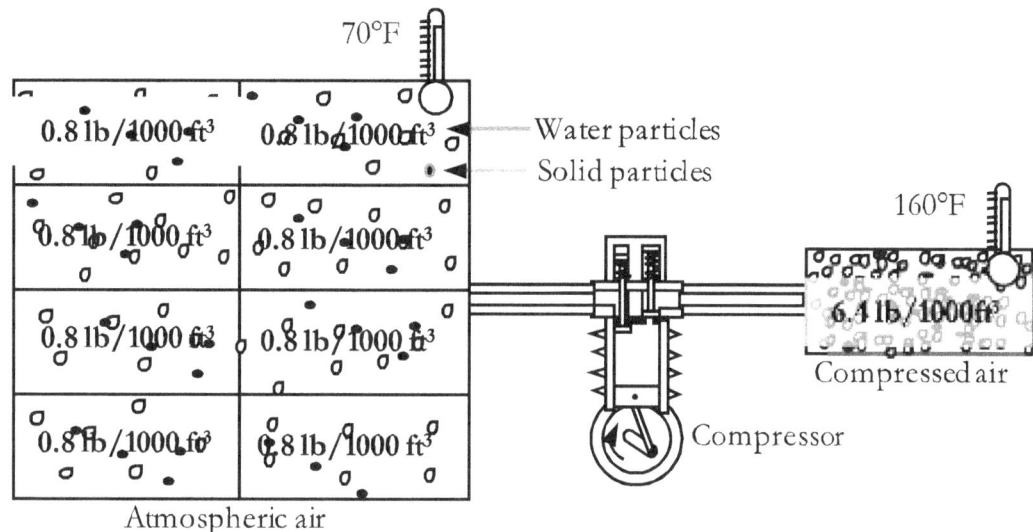

Figure 4.2 | Condition of air at the outlet of a compressor

The condition of the compressed air at the compressor outlet is shown in Figure 4.2.

Effects of Contamination

The presence of contaminants can have a detrimental impact on pneumatic systems. Water droplets resulting from the condensation can cause rusting of exposed surfaces, the formation of sticky emulsions, and consequent jamming of valves. Corrosion caused by these contaminants can shorten the lifespan of equipment and potentially compromise product quality in the food and beverage manufacturing industry, as well as the health of employees. Additionally, the presence of oil can negatively affect the performance and finish of paint spraying, powder coating, and powder conveying. Table 4.2 gives the contaminants in pneumatic systems and their effects.

Table 4.2 | Contaminants in pneumatic systems and their effects

Contaminant	Effects
Dirt	cause blockages and affect production quality
Rust	cause blockages and damage to seals and equipment
Water	cause increased leakage, corrosion, reduced tool performance, high maintenance costs, disturbances in control systems, bacterial growth, etc
Oil	negatively affect plant equipment and products that come in contact with oil-contaminated air

In general, contamination harms the performance of pneumatic systems. This can ultimately lead to an increase in maintenance and repair costs and a decrease in the equipment's lifespan.

Stages of Compressed Air Preparation

To achieve any degree of reliability, the components of pneumatic systems must get clean and dry air. Hence, air must be prepared or conditioned before it can enter a pneumatic system. Compressed air preparation consists of reducing its temperature, removing water and solids from it, regulating its pressure, and in many cases introducing lubricant in it. In general, the preparation of air falls into three distinct stages, as shown in Figure 4.3.

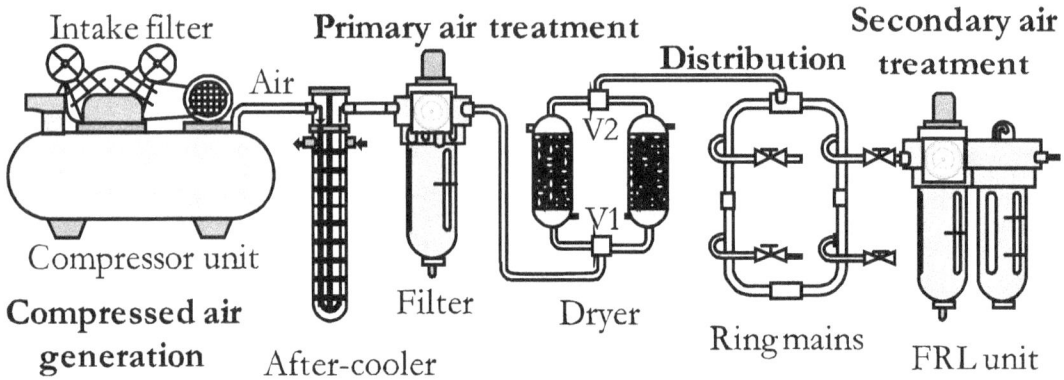

Figure 4.3 | Stages of compressed air preparation

Intake Filter

The intake filter removes large particles, usually of sizes greater than 200 microns, which can damage the air compressor.

Compressor and Storage Unit

A compressor generates compressed air. A receiver tank stores the compressed air.

Primary Air Treatment

Primary air treatment is intended to reduce the temperature of the air at the outlet of the compressor, remove solid contaminants, usually in the range from 5 to 100 microns, present in it, and dry the air to reduce its humidity.

The units used in the primary air treatment are the aftercooler, main-line filter, and dryer.

Aftercoolers

Typically, the temperature of air exiting a compressor is between 180°F to 350°F. An aftercooler is intended to reduce the temperature of compressed air discharged by a compressor to approximately 5 to 30°F over the temperature of the cooling medium. Typically, aftercoolers are designed to achieve an approach temperature of 5, 10, 15, 20, or 30°F above the cooling-medium temperature.

A stand-alone aftercooler is a separate unit installed downstream of the compressor. Compressor manufacturers may include aftercoolers within the compressor package.

Types of Aftercoolers

The two basic aftercoolers are: (1) Air-cooled aftercoolers and (2) Water-cooled aftercoolers.

Air-cooled Aftercooler

Air-cooled aftercoolers use ambient air to cool the hot compressed air. Figure 4.4 gives the essential parts of an air-cooled aftercooler. The compressed air travels through finned tubes while a motor-driven fan forces ambient air over the cooler. The forced air removes heat from the compressed air. As the air cools, the moisture in the compressed air condenses. When the air reaches the cooler's separator, centrifugal motion causes the condensed water and other contaminants to hit the cylinder walls and drip down the drain.

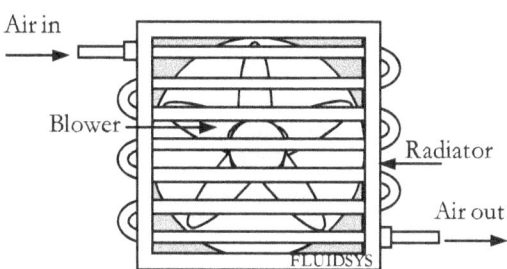

Figure 4.4 | Air-cooled aftercooler

Water-cooled Aftercooler

The standard style is the shell and tube aftercooler, in which a bundle of copper tubes is fitted inside the shell, as shown in Figure 4.5. The hot compressed air flows through the tubes in one direction while the cooling water flows in the opposite direction around the tubes in the shell. As the compressed air is cooled, moisture condenses out of the air in the form of water. The moisture separator and drain valve remove the water.

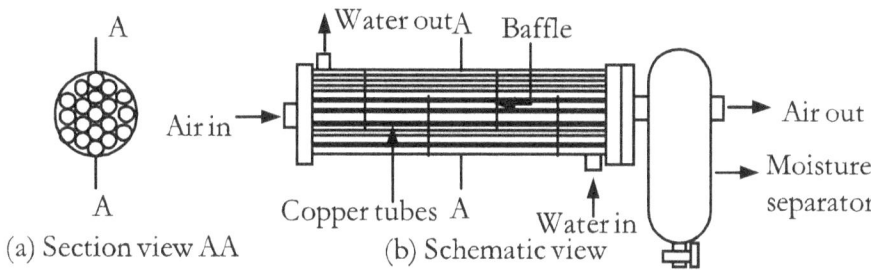

Figure 4.5 | Water-cooled aftercooler

The tube bundles can be fixed or removable. Fixed tube bundles cost less but are more difficult to maintain than removable bundles that can be detached for cleaning or service.

Condensate Separator, Aftercooler

It will remove the liquid condensate. It is installed at the base of the aftercooler.

Automatic Condensate Drains, Aftercooler

Condensate drips down the walls of the separator into the automatic condensate drain. As the drain fills, a float rises, opening a valve that empties the condensate. This automatic action ensures that condensate does not build up in the cooler.

Dryers are generally designed for a specific inlet temperature. If the system experiences a higher temperature, the dryer must be oversized. Table 4.3 gives typical specifications for air-cooled aftercoolers, and Table 4.4 gives typical specifications for water-cooled aftercoolers.

Aftercooler Sizing Considerations

First, determine the compressor outlet temperature. The temperature at the outlet of a two-stage reciprocating piston compressor is typically about 250°F, and that for a rotary screw compressor is about 200°F. Next, find the maximum free air delivery in terms of cfm and the pressure in terms of psi.

After determining the parameters, a correctly-sized aftercooler can be selected from the manufacturer's size charts.

Typical Specifications of Air-cooled Aftercoolers

Table 4.3 | Specifications of air-cooled aftercoolers

Capacity cfm	Max working pressure psi	Max air inlet temp (°F)	Approach temperature (°F)	Electric supply (VAC)	In/out ports BSP
35	230	250	$t_{amb} + 15°F$	220	½"
50	230	250	$t_{amb} + 15°F$	220	¾"
100	230	250	$t_{amb} + 15°F$	220	1½"
150	230	250	$t_{amb} + 15°F$	220	1½"
200	230	250	$t_{amb} + 15°F$	220	2"NB flg
300	230	250	$t_{amb} + 15°F$	220	2"NB flg
400	230	250	$t_{amb} + 15°F$	415	2"NB flg
600	230	250	$t_{amb} + 15°F$	415	2"NB flg
750	230	250	$t_{amb} + 15°F$	415	2"NB flg
1000	230	250	$t_{amb} + 15°F$	415	2"NB flg

t_{amb} = Ambient temperature, °C

Typical Specifications of Water-cooled Aftercoolers

Table 4.4 | Specifications of water-cooled aftercoolers

Capacity cfm	Max working pressure psi	Max air inlet temp (°F)	Approach temperature (°F)	Air in/out ports BSP	Water in/out ports BSP
50	230	300	$t_{amb} + 15°F$	¾"	½"
100	230	300	$t_{amb} + 15°F$	1"	½"
200	230	300	$t_{amb} + 15°F$	1½"	¾"
250	230	300	$t_{amb} + 15°F$	2"NB flg	1"
400	230	300	$t_{amb} + 15°F$	3"NB flg	1½"
500	230	300	$t_{amb} + 15°F$	3"NB flg	1½"
750	230	300	$t_{amb} + 15°F$	3"NB flg	2"
100	230	300	$t_{amb} + 15°F$	4"NB flg	2"
1250	230	300	$t_{amb} + 15°F$	4"NB flg	2"
1500	230	300	$t_{amb} + 15°F$	5"NB flg	2"

t_{amb} = Ambient temperature, °C

Compressed Air Filters

Filters can be fitted to the mainline of a compressed air system to remove dust, dirt, oil, and water from the compressed air. They can clean the air to a recognized compressed air purity standard, such as ISO 8573. They are often used with refrigerant and desiccant-type dryers. A schematic diagram of a pneumatic mainline filter is shown in Figure 4.6.

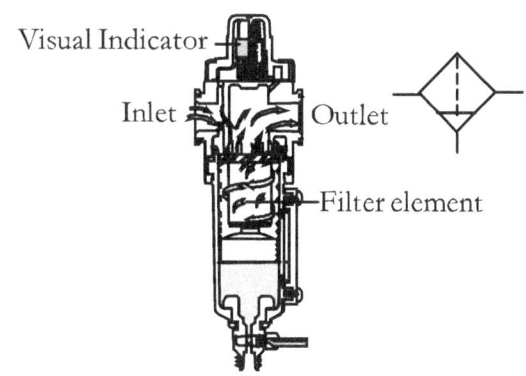

Figure 4.6 | Mainline filter

Types of Filters

Pneumatic filters can generally be classified as: (1) General purpose filters, (2) Coalescing filters, and (3) Adsorbing filters.

General-purpose filters can remove solid particles down to 5 microns and water droplets from the system.

Coalescing filters can remove 99.99% of oil contaminants and solids greater than 0.3 microns in size. It can provide air quality 1.7.2 as per the ISO standard 8573-1. However, it cannot remove oil vapors.

An adsorbing filter is an ultra-high efficient coalescing filter with an active carbon pack. The filter media can attract and remove oil vapors, and the carbon pack can remove hydrocarbon gases. It can provide an air quality of 1.7.1 as per the ISO standard 8573-1.

Application Notes on Filters

Filters have different capabilities based on their media. Particulate filters remove larger particles like dirt and rust, while finer filters can remove water and oil aerosols. These filters can be used with lubricated or non-lubricated compressors.

Coalescing-type oil removal filters direct air through a maze of submicronic glass fibers, where oil aerosols are coalesced into larger droplets and removed. These filters effectively remove oil aerosols that contaminate end products, spoil paint jobs, and damage air tools.

Oil vapor adsorbers are designed to eliminate smell and taste by removing oil vapors and other hydrocarbons. They are installed after the oil removal filter to prevent premature saturation of the activated carbon by liquid oil aerosols, which could significantly reduce adsorptive capacity.

Typical Specifications of Filters

Table 4.5, Table 4.6, and Table 4.7 give typical specifications for different types of filters. [1 l/s = 2.11888 scfm]

Table 4.5 | Sample specifications of general-purpose filters

Capacity (scfm)	Working pressure (psi)	Element size* (μm)	Connection size (BSP)
70	230	5, 20, 40	G¼
140	230	5, 20, 40	G⅜
160	230	5, 20, 40	G½
160	230	5, 20, 40	G¾
340	230	5, 20, 40	G¾
400	230	5, 20, 40	G1
425	230	5, 20, 40	G1¼
425	230	5, 20, 40	G1½

*Filter element material: Polyethylene – 5-μm-rated; Sintered bronze – 5-μm, 20-μm, or 40-μm –rated.

Table 4.6 | Sample specifications of coalescing filters

Capacity (scfm)	Working pressure (psi)	Element size (μm)	Connection size (BSP)
34	230	0.01	G¼
60	230	0.01	G⅜
60	230	0.01	G½
60	230	0.01	G¾
74	230	0.01	G½
74	230	0.01	G¾
127	230	0.01	G1

Table 4.7 | Sample specifications of adsorbing filters

Capacity (scfm)	Working pressure (psi)	Element size (μm)	Connection size (BSP)
14.8	230	0.01	G¼
23.3	230	0.01	G⅜
23.3	230	0.01	G½
23.3	230	0.01	G¾
53	230	0.01	G½
74.2	230	0.01	G¾
127	230	0.01	G1

Drain

A drain mechanism can be provided in filters to remove the water before it re-enters into the downstream air. The mechanism can be of the manual type or automatic type. The standard drain is a manual type. The auto drain facility ensures that the filter bowl is drained without operator intervention. There are two types of automatic drains. They are: (1) Float type and (2) Differential pressure type.

Float-type drain works on the float principle. As water accumulates in the bowl, the float will be lifted. As a result, a passage will open in the bowl. The air pressure in the bowl will then vent to the atmosphere through the opening, blowing the water away from the bowl. When the water is completely drained, the float drops back into the orifice, sealing off the passage in the bowl. If an application is continuous, a float-type drain must be used in a filter used in the application.

There must be a pressure differential across the filter in the differential pressure design.

An external drain is used where severe condensation problem exists. The water will be removed with the self-flushing action of the filter.

Additional Specifications of Filters

Filter bowl:
- Transparent (Polycarbonate) with guard
- Metal with liquid level indicator

Service life indicator: Mechanical | Electrical

Drain: Manual | Auto drain | External

Threads: PTF | ISO taper | ISO parallel

Dryers

The natural water vapor content of air is concentrated and is carried through the compression process as a vapor in high temperatures. All that may be essential for simple applications is an aftercooler, air receiver, and filter with condensate traps to remove the excess humidity.

Additional dehydration means must be provided where the demand for high-quality compressed air is entailed. The most commonly used compressed air drying methods are: (1) Adsorption drying and (2) Refrigeration drying.

A dryer can optionally be provided with a pre-filter and a post-filter for optimum performance. The pre-filter is typically a coalescing filter of submicron (0.1 to 0.01 μ) level. The protect the pre-filter, another filter of 1 to 5 μ level is usually provided. The pre-filter can prevent oil carryover from the compressor from entering the dryer and contaminating the drying material.

The post-filter can be a particulate filter of one-micron level or an adsorbing filter. The particulate filter downstream of the dryer may be used to protect downstream equipment from desiccant dust. An adsorbing filter with an active carbon pack may be used to attract and remove oil vapors and hydrocarbon vapors to provide air quality 1.7.1 as per the ISO standard 8573-1.

Using an aftercooler before any air dryer is advisable to reduce the work enforced on the dryer. An air dryer is ideally fitted downstream of the compressor and reservoir.

Adsorption Dryer

Adsorption is the physical process of collecting moisture on the porous surface of certain granular materials such as silicon dioxide (Silica gel), activated alumina, and copper sulphate. Figure 4.7 shows the constructional features of a typical adsorption dryer. When compressed air is passed through the drying agent, the moisture present in the air is adsorbed by the drying agent until it gets saturated. Dry compressed air is delivered out of the dryer.

The silica gel drying agent changes color as it approaches the saturation point. When saturated, the drying agent can be renewed by blowing warm or cold air through the material, which then takes up the moisture. The silica gel reassumes its original color when the moisture is driven out. This method of drying is also known as regenerative drying.

In practice, two parallel chambers are used for non-stop production. While one chamber is drying the air, the other can be set for regeneration.

Types of Adsorption Dryers
Adsorption dryers can be of the following three types: (1) Heatless type, (2) Heated type, and (3) Heated blower type.

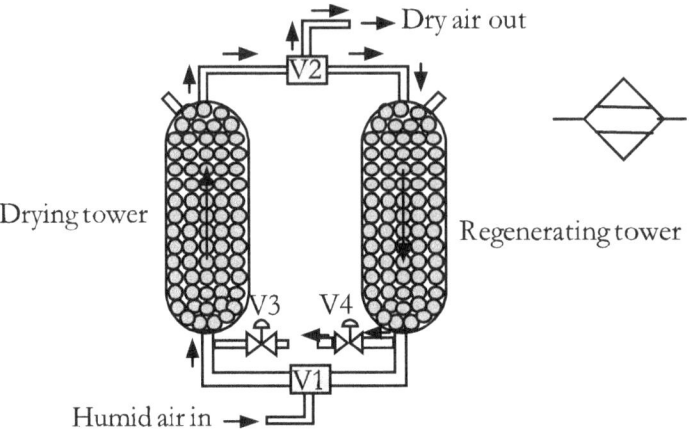

Figure 4.7 | Adsorption dryer

- Heatless twin tower dryer diverts some dried air to the off-line tower. This dry air then flows through the saturated desiccant and regenerates it.

- In the heated type dryer, a portion of the dried air is first passed through a high-efficiency external heater before entering the off-line tower to regenerate the desiccant.

- Heated blower type dryer employs a high-performance centrifugal blower to direct ambient air through a heater and the off-line tower. The stream of heated air then regenerates the desiccant.

Refrigerated Dryer

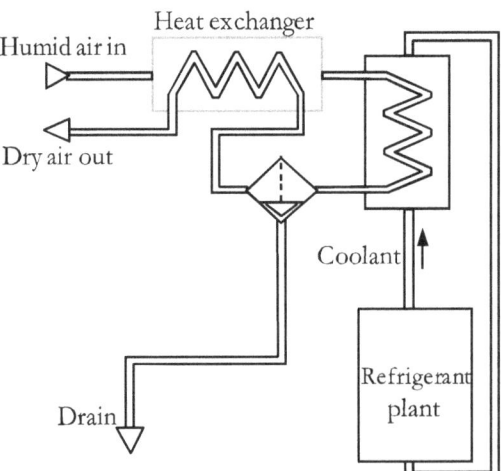

Figure 4.8 | Low-temperature dryer

The schematic of a typical refrigerated air dryer is shown in Figure 4.8. It consists of a heat exchanger and a refrigerating unit. In the first stage, the warm and humid compressed air is passed through the heat exchanger. The air gets cooled to the near-ambient temperature condition of the heat exchanger. The moisture present in the air gets condensed, corresponding to the temperature, and water is precipitated.

In the second stage, the partly prepared air is passed through the refrigerating unit to reduce the compressed air temperature to as low as 35°F. Again the moisture is condensed corresponding to the temperature in the refrigerating unit. The condensed water can then be collected in the water traps provided at appropriate points. Finally, the air again goes through the heat exchanger and gets discharged clean and dry.

Pressure Dew Point

The pressure dew point is the lowest air temperature reached during the drying process at the specified operating pressure. Adsorption dryers can typically reach pressure dew points of -94°F, -40°F, and -4°F, while refrigerated dryers can achieve ++37°F, +45°F, and +50°F.

Selection of Dryers

The selection of dryers depends on the variables, such as system demand, compressed air capacity, air quality requirements, and applicable life cycle costs unique to the compressed air system.

Tables 4.8 to 4.12 give sample specifications for various types of dryers.

Typical Specifications of Desiccant Type Dryers

Table 4.8 | Typical specifications of basic desiccant-type dryers

Capacity (scfm)	Pressure dew point, °F	Pressure rating (psi)	Connection (in)
90	-40	150	1 NPT
120	-40	150	1 NPT
160	-40	150	1½ NPT
200	-40	150	1½ NPT
250	-40	150	1½ NPT
300	-40	150	2 NPT
400	-40	150	2 NPT
500	-40	150	2 NPT
600	-40	150	2 NPT
800	-40	150	3 NPT
1000	-40	150	3 NPT
1200	-40	150	3 NPT
1500	-40	150	4 FLG
1800	-40	150	4 FLG
2100	-40	150	4 FLG
2700	-40	150	4 FLG
3300	-40	150	6 FLG
4000	-40	150	6 FLG
5000	-40	150	6 FLG

Typical Specifications of Heated Desiccant Type Dryers

Table 4.9 | Typical specifications heated desiccant type dryers

Capacity (scfm)	Pressure dew point, °F	Heater rating of the heated dryer, if used (hp)	Pressure rating (psi)	Connection (in)
150	-40	2.7	150	1 NPT
200	-40	4	150	1½ NPT
250	-40	4	150	1½ NPT
300	-40	4	150	1½ NPT
400	-40	6	150	2 NPT
500	-40	6	150	2 NPT
600	-40	8	150	3 NPT
800	-40	12	150	3 NPT
1000	-40	12	150	3 NPT
1200	-40	16	150	3 NPT
1500	-40	20	150	3 NPT
1800	-40	24	150	4 FLG
2100	-40	24	150	4 FLG
3000	-40	40	150	4 FLG
4000	-40	48	150	6 FLG
5000	-40	67	150	6 FLG
6000	-40	80	150	6 FLG

Typical Specifications of Heated Blower Type Desiccant Dryers

Table 4.10 | Typical specifications heated blower type desiccant dryers

Capacity (scfm)	Heater rating of the heated dryer, if used (hp)	Blower rating of the blower, if used (hp)	Connection (in)
150	4	1	1 NPT
200	6	1	1½ NPT
250	8	1.5	1½ NPT
300	8	1.5	1½ NPT
400	12	2	2 NPT
500	16	2	2 NPT
600	16	5	3 NPT
800	24	5	3 NPT
1000	32	7.5	3 NPT
1200	32	7.5	3 NPT
1500	40	15	3 NPT
1800	48	15	4 FLG
2100	60	15	4 FLG
3000	80	20	6 FLG
4000	107	25	6 FLG
5000	134	30	6 FLG
6000	168	30	6 FLG

Pressure dew point – 40°F | Pressure rating – 150 psi

Typical Specifications of Refrigeration Dryers

Table 4.11 | Specifications of refrigeration dryers
(230V, 1Ph, 50 Hz)

Capacity (scfm)	Electrical power, hp	Pressure (psi)	Connection (in)
20	0.4	44 to 230	G ¾
30	0.4	44 to 230	G ¾
45	0.4	44 to 230	G ¾
75	0.75	44 to 230	G 1
90	0.85	44 to 230	G 1
110	1	44 to 230	G 1¼
140	1.2	44 to 230	G 1¼
160	1.5	44 to 230	G 1¼
200	1.15	44 to 230	G 1½
250	1.5	44 to 230	G 1½
300	1.8	44 to 230	G 2

Table 4.12 | Specifications of refrigeration dryers
(400V, 3Ph, 50Hz)

Capacity (scfm)	Electrical power, hp	Pressure (psi)	Connection (in)
400	1.5	100	G 2
440	1.5	100	G 2
510	2	100	G 2
600	1.8	100	DN 65
810	2.6	100	DN 80
1000	3.3	100	DN 80
1200	3.6	100	DN 80
1600	4.4	100	DN 100
1835	5.2	100	DN 100
2300	6.5	100	DN 150
2750	7.9	100	DN 150
3460	13	100	DN 150

Pressure dew point +35°F | Inlet air temperature +140°F

Typical Air Treatment Configurations

The quality of compressed air needed depends on the specific application. Suitable filters and dryers can be used to achieve the necessary level of quality. While some applications require the highest quality compressed air, others may have less strict requirements. However, it is important to note that preparing compressed air to a high quality can be expensive. Thus, the air preparation equipment must be chosen carefully to ensure that compressed air is prepared economically to the appropriate quality level. A table (Table 4.13) outlines various application-specific quality classes and filter/dryer configurations to achieve those levels.

Table 4.13 | Typical compressed air quality classes and configurations

Quality Class*	Grades of filtration (µm)	Pressure dew point	Applications
1.1.1	5 + 1 + 0.01 + activated carbon	-100°F	-Breweries / Dairies /Electronics
1.2.1	5 + 1 + 0.01 + activated carbon	-40°F	-Pharmaceutical and Food [Direct contact between air and dry products] -Electronics (Chip and Data disc production)
1.3.1	5 + 1 + 0.01 + activated carbon	-4°F	-Semiconductor -Pharmaceutical -Measurement and Test Air (3D measurement technology) -Surface Finishing (Powder coating, painting)
1.4.1	5 + 1 + 0.01 + activated carbon	37°F	-Pharmaceutical and Food [Direct contact between air and non-dry products/packaging material] -Electrical/Electronics (Lighting and CD manufacturing) -Surface Finishing (Powder coating, painting) -Chemical Industries -Chemical Fiber Production -Rubber and Plastic Industry (Conveying air) -Measuring and Testing Systems
1.4.2	5 + 1	37°F	-Instrument Air / Paint Spraying / Powder Coating
2.4.4	5 + 1	37°F	-Air Tools -Sand Blasting
3.4.2	5 + 0.01	37°F	-Textile (Conveying air) -Paper (Conveying air) -Publishing and printing (Conveying air) -Glassworks and Ceramics (Conveying air) -Rubber and plastic industry (Blast air)
5.4.3	5 + 1	37°F	-Metal Production and Machining -Foundries -Machine and Plant Construction (Blast air)
7.4.4	40	37°F	-Pilot Air, General

*Quality Classes to ISO 8573-1:2010

-Pilot air does not come into direct contact with the products. It operates valves, cylinders, and grippers
-Blast air is used to clean machines and workpieces.
-Process air is involved in machining or processing.

Chapter 5 | Compressed Air Distribution Systems

The objective of the air distribution system is to act as a leak-proof carrier of compressed air and limit pressure drops within permissible limits. The air distribution system comprises conductors and fittings, interconnecting various pneumatic system components. A typical pneumatic distribution layout is shown in Figure 5.1. A well-organized industrial pneumatic distribution system is designed with correctly sized pipes and components, ensuring minimum elbows and bends so that pressure energy is not unnecessarily wasted. Distribution of compressed air should be planned and executed carefully by taking into account the following considerations: (1) correct sizing of pipes and fittings, (2) choice of pipe materials, (3) pipe layout, and (4) the total cost of the conductor system.

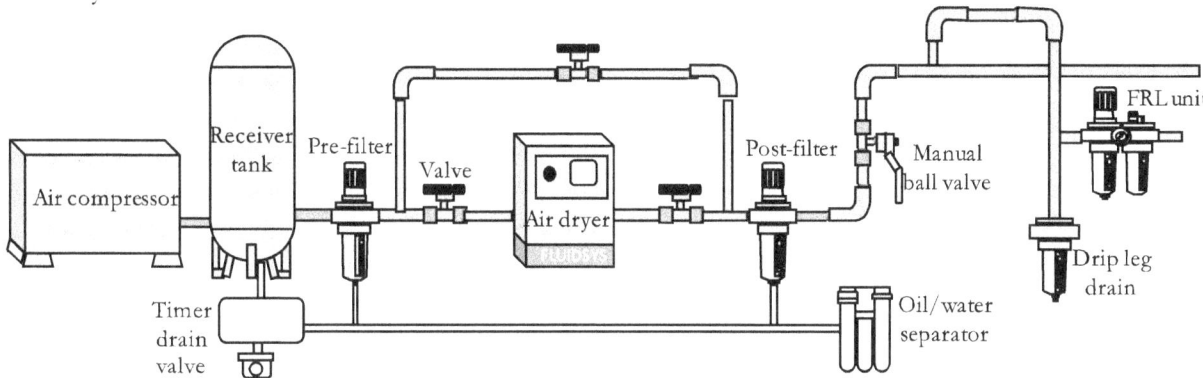

Figure 5.1 | A typical pneumatic layout

Conductors
The conductors are generally divided into three classes: (1) Pipe (Rigid), (2) Tubing (Rigid or semi-rigid), and (3) Hose (Flexible). More than one type of conductor may be used in the same installation.

Rigid pipe
The main distribution system comprises rigid pipelines, feeder lines, associated fittings, and accessories. Copper, iron, steel, and aluminum pipes must be brazed and welded or can be joined using threaded connectors. Welded connections are robust, leak-free, and the primary choice for fixed main distribution pipelines. As a rough rule, piping is employed for diameters above 2 inches. Hoses and tubing conduct compressed air to air-powered tools and equipment, instruments, and gauges.

Tubing
Semi-rigid type and flexible type of tubing are available for use in pneumatic applications. Examples of rigid types of tubing are steel, aluminum, copper, and polyvinyl chloride, and flexible type are nylon and polyethylene. Nylon tubes are robust and can be used for various applications within general pneumatics. Polyurethane tubes are extra flexible and soft and are especially suitable in applications where the requirement of short bending radii for tubing is indispensable.

Each material has definite characteristics, which make it more appropriate for some services than others. Since tubing can be bent, lines from tubing require only a minimum number of fittings. A tube is usually specified by outside diameter and wall thickness.

Plastic tubing is flexible and has gained full acceptance in the industry as conductors in pneumatic systems, as lengths of tubing are inexpensive and extremely easy to use with a high degree of flexibility. Food-grade tubes are colorless and tasteless and will not pass on extraneous flavor or odor to susceptible foods or beverages.

Hose

Hose assemblies connect compressed air sources to actuators that must be located on movable parts or because of the need to bend lines. The advantages of using hoses are that they can be easily installed, require fewer installation skills than required for pipes or rigid tubing, absorb shock, and are readily available in various pressure ratings.

Flexible hoses are manufactured from natural and synthetic rubbers and several plastics and are reinforced by fabric or wire braid. Examples are: (1) polyester-reinforced PVC hose and (2) metal braided rubber hose. A hose is usually specified by the inside and outside diameters. A hose must have a smooth bore and resist oil vapors and lubricants. The wall of the hose must be sufficiently hard to resist heavy impacts and shock blows. The outer structure of the hose must be strong and abrasive-resistant.

Fittings

Pipes and tubes are joined to other pipes and tubes or the components of an installation by using some connectors. Remember, there are many different types of connectors available for pneumatic systems. Some examples of fittings are push-in, push-on, and compression fittings. Push-in fittings are used for quick and straightforward assembly of pneumatic circuits. They are compact units comprising retained collets and positive tube anchorage for easy insertion and rapid assembly. Fittings are made of stainless steel, aluminum, bronze, and plastic with silicon-free nitrile rubber / Viton 'O' rings. They are available in various shapes to form unions, elbows, tees, nipples, caps, plugs, couplings, and crosses.

Quick-disconnect coupling

Quick-disconnect couplings are widely used in pneumatic systems, mainly where there are frequent needs to uncouple the lines for maintenance, testing, and safety. Many disconnect couplings have double checks that can be used for easy detachment without losing compressed air.

Air Fuse

In a pneumatic system, the energy stored will be expelled quickly with great velocity and force when a flexible tubing or hose breaks or a tool attached to the flexible line is unexpectedly uncoupled. This sudden surge of compressed air through the open end of the line can cause it to whip around uncontrollably, which can be extremely dangerous and potentially life-threatening. To prevent such accidents, pneumatic systems should have air fuses that automatically shut off airflow if a line is severed. The air fuse should be installed directly between the rigid pipework and the flexible line to protect the whole length of the flexible line—only the line after the air fuse is protected. The air fuse must be installed in the correct direction for airflow.

Pipe threads

Threaded pipe connections must contain male threads on the pipes. Threads are available to a variety of standards, some of which are: American National Pipe Threads (NPT), Unified Pipe Threads (UNF), British Standard Pipe Threads (BSP), and Metric Pipe Threads (M). The choice between these standards is determined by those already chosen for a user's region or country. Taper threads are cone-shaped and form a seal between the male and female parts as they tighten, with assistance from some jointing compound or plastic tapes.

Flow resistance

The flow of compressed air through piping creates friction and consequent pressure drop. It may be noted that the pressure loss is proportional to the square of the velocity of the flow. Elbows, T-pieces, two-way valves, and slide valves also interfere with the flow and the corresponding pressure loss. However, this pressure drop cannot be avoided but can be considerably reduced by routing pipes properly and assembling the fittings correctly.

Pipe layout

Depending on usage requirements, plant size, and delivery volume, various piping arrangements can be used in air distribution systems. Generally, the distribution is arranged as a manifold, as shown in Figure 5.2(a), or as a ring main, as shown in Figure 5.2(b).

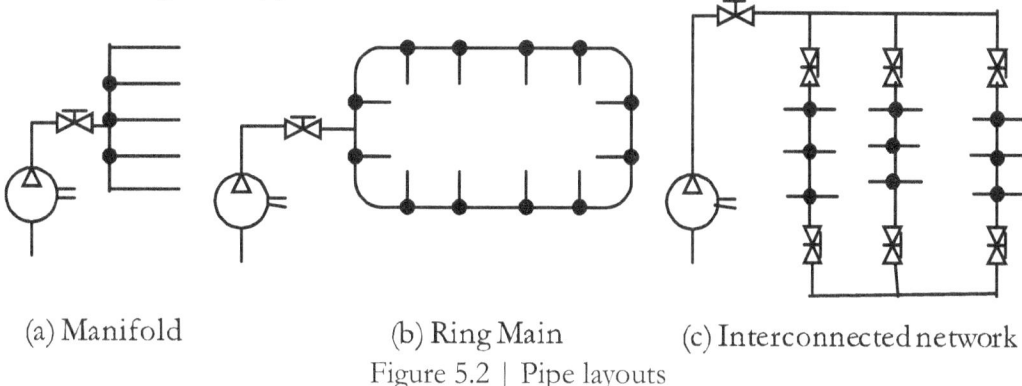

(a) Manifold   (b) Ring Main   (c) Interconnected network

Figure 5.2 | Pipe layouts

As the actuating devices consume air, the pressure is decreased downstream. One technique for compensating for the pressure drop is to use the ring-main layout. This layout can meet any demand for compressed air in two directions. A ring main ensures largely uniform pressure conditions in the air network.

With an interconnected network system, as shown in Figure 5.2(c), parts of the ring can be separated using the shut-off valves to maintain, repair, and extend the network without disturbing the rest of the system.

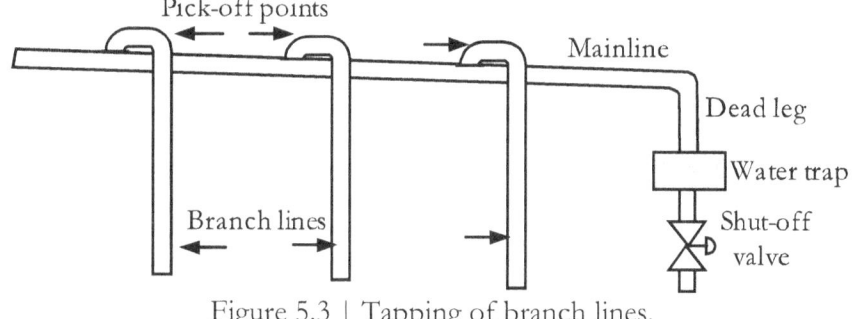

Figure 5.3 | Tapping of branch lines.

The distribution pipe system can be considered part of the storage, and the compressed air inside is also subjected to external cooling. This cooling causes the moisture in the air to condense and consequently precipitate water. Hence, to provide drainage, the pipes should be inclined 1 to 2% downward, in the airflow direction, preferably to each corner.

Additionally, all take-off points are tapped from the top of the pipe, as shown in Figure 5.3, to prevent water entry into the branch lines. The condensate can then be released from the system through a dead leg at the lowest point.

An automatic drain valve can be provided for terminating a dead leg. Accumulated water can then be automatically drained off when pressure is on and the system is shut down.

Heavy demands for compressed air are to be met occasionally at the ends of long lines, which can result in severe pressure loss.

Installing intermediate reservoirs close to the demand points can avoid this pressure loss.

Typical Tubing Specifications

Table 5.1 and Table 5.2 give typical specifications for nylon tubing and polyurethane tubing, respectively.

Table 5.1 | Specifications of nylon tubing

OD (in)	ID (in)	Maximum pressure (psi) @150 °F	Maximum pressure (psi) @150 °F	Minimum bend radius (in)
1/8	0.079	270	160	0.75
3/16	0.138	260	160	1.5
1/4	0.125	260	160	1.75
5/16	0.232	250	140	2.0
3/8	0.275	250	140	2.5
1/2	0.375	240	140	3.0
5/8	0.5	240	140	4.0
3/4	0.6	240	140	6.0

Table 5.2 | Specifications of polyurethane tubing

OD (in)	ID (in)	Maximum pressure (psi) @75°F	Maximum pressure (psi) @150°F	Minimum bend radius (in)
1/8	1/16	255	100	1/4
5/32	3/32	210	85	3/8
3/16	0.107	220	90	3/8
1/4	1/8	265	105	1/2
3/8	1/4	175	70	7/8
15/32	5/16	155	60	1 1/8
1/2	0.32	175	70	1 1/8
9/16	3/8	155	60	1 3/8
3/4	0.467	175	70	1 1/2

Sizing of pipe systems

An under-sized pipe in a pneumatic system produces a significant pressure drop and consequent energy losses. The pipe size should be selected appropriately to keep the pressure constant throughout the system. An oversized pipe costs more. The correct sizing of each part of the pipe system should be ensured for the air distribution system's reliable, efficient, and trouble-free operation.

The following factors govern the selection of pipe size: (1) Delivery volume of the compressor, (2) Total pipe length, (3) Operating pressure, (4) Permissible pressure drop (maximum 1 bar), (5) Use of fittings such as elbows, T-pieces, and valves.

Pipe manufacturers provide charts or nomograms linking delivery volumes, pipe lengths, operating pressures, and permissible pressure drops to different pipe diameters. A typical chart for finding pipe sizes is given in Table 5.3.

Steps for Finding Pipe Size

1. Determine to find the maximum delivery of the compressor system in scfm

2. Draw the piping schematic with fittings and valves

3. Mark the lengths of straight pipes on the schematic and determine the total length of straight pipes

4. Find the initial pipe size corresponding to the scfm and the total length of straight pipes from Table 5.3

5. Next, find the equivalent length of each pipe fitting corresponding to the pipe size determined from the previous step from Table 5.4

6. Find the total of the equivalent lengths of all fittings

7. Add the total length of straight pipes and equivalent lengths of all fittings to get a revised pipe length

8. Find the pipe size corresponding to the revised pipe length from Table 5.3

Compressed Air Pipe Sizing Chart for Straight Pipes

Table 5.3 is a pipe sizing chart for straight pipes. The pipe sizes are listed in the Table, assuming a 100-psi pneumatic system to carry air at a 1-psi loss per 100 feet. The hp rating given is for the associated compressor. For pressures other than 100 psi, the flow rate can be calculated as per the Boyles law.

Table 5.3 | Pipe diameter in inches

scfm	Length of pipe run, ft									HP
	25	50	75	100	150	200	300	500	1000	
4	½	½	½	½	½	½	½	¾	¾	1
12	½	½	½	¾	¾	¾	¾	1	1	3
20	¾	¾	¾	¾	1	1	1	1¼	1¼	5
30	¾	¾	1	1	1	1	1¼	1¼	1¼	7½
40	¾	1	1	1	1¼	1¼	1¼	1½	1½	10
60	1	1	1¼	1¼	1¼	1¼	1½	1½	2	15
80	1	1¼	1¼	1¼	1½	1½	1½	2	2	20
100	1¼	1¼	1½	1½	1½	1½	2	2	2½	25
120	1¼	1½	1½	1½	2	2	2	2½	2½	30
160	1¼	1½	1½	2	2	2	2½	2½	3	40
200	1½	2	2	2	2	2	2½	3	3	50
240	1½	2	2	2	2½	2½	2½	3	3	60
300	2	2	2	2½	2½	2½	3	3	3½	75
400	2	2½	2½	2½	3	3	3	3½	4	100
500	2	2½	2½	3	3	3	3½	3½	4	125

Equivalent Length of Pipes for Fittings

Bends, couplings, and other restrictions also increase the pressure drops. The pressure drops in pipe fittings are generally specified in terms of equivalent lengths of a standard pipe. Typical equivalent pipe lengths are given in Table 5.4.

Table 5.4 | Equivalent pipe lengths for fittings in ft

Fitting	Pipe diameter (inch)						
	½	1	1½	2	2½	3	4
Elbow, 90° nominal	1.5	2.6	4.0	5.7	6.9	7.9	11.4
Elbow, 45° standard	0.8	1.4	2.1	2.7	3.3	4.1	5.3
Standard Tee run flow	1.0	1.7	2.7	4.3	5.1	6.2	8.3
Standard Tee branch flow	4.0	6.0	8.1	12.0	14.7	16.3	22.0
Two-way valve	2.5	4.7	7.8	10.6	13.1	17.1	23.7

An example of finding the pipe diameter using charts is illustrated in Example 5.1.

Example 5.1

The air consumption in an industrial plant is found to be 100 scfm. Next, the likely increase in air consumption for about three years is 2 times the current air consumption. The total length of straight pipes is measured to be 280 ft. Additionally, the distribution network contains T-pieces (4 Nos.), normal elbows, 90° (3 Nos.), and a two-way valve (1 No.). The permissible pressure drop is limited to 14.5 psi, and the operating pressure is 100 psi. Calculate the pipe diameter using the nomograms (Figures 3.12 and 3.13).

Given

Air consumption, present	= 100 scfm
Air consumption, expected increase (In three years)	= 2 x 100 = 200 scfm
Total air consumption, (Considering future expansion)	= (100 + 200) scfm = 300 scfm
Total length of straight pipes	= 200 ft
Pressure	= 100 psi
Pressure drop	= 14.5 psi

Referring to the chart for sizes for straight pipes given in Table 5.3, find the initial pipe size corresponding to the scfm and the total length of straight pipes.

Pipe size (Initial calculation) = 2½ inch

The equivalent pipe lengths for fittings are found in Table 5.4 in the following manner:

3 elbow pieces	= 3 x 6.9 ft = 20.7 ft
4 T-pieces, ranch flow	= 4 x 14.7 ft = 58.8 ft
1 two-way valve	= 1 x 13.1 ft = 13.1 ft
Total equivalent lengths	= 92.6 ft ~ 100 ft
Total pipe length	= Total length of straight pipes + Total equivalent lengths

= (200+100) ft = 300 ft

With this modified pipe length, find the pipe diameter again using the chart for pipe diameter.

The revised pipe diameter = 3 inch

Calculation of Internal Pipe Diameter, Alternative Method

The internal pipe diameter (d) can be calculated using the following equation:

$$d = \sqrt[5]{\frac{283 \times Q^{1.85} \times l}{P \times \Delta P}}$$

Where,

> d = Internal pipe diameter, in
> Q = Free air delivery, compressor (cfm)
> l = overall pipe length, (ft)
> P = Working pressure [psi(a)]
> ΔP = Permissible pressure drop (psi)

Example 5.1 | Calculation of Internal Pipe Diameter Using Mathematical Relation

The air consumption in an industrial plant is found to be 100 scfm. Next, the likely increase in air consumption for about three years is 2 times the current air consumption. The total length of straight pipes is measured to be 280 ft. Additionally, the distribution network contains T-pieces (4 Nos.), normal elbows, 90° (3 Nos.), and a two-way valve (1 No.). The permissible pressure drop is limited to 14.5 psi, and the operating pressure is 100 psi. Calculate the pipe diameter using the nomograms (Figures 3.12 and 3.13).

Given

> Air consumption, present   = 100 scfm
> Air consumption, expected increase = 2 x 100 = 200 scfm
> (In three years)
> Total air consumption,   = (100 + 200) scfm
> (Considering future expansion)   = 300 scfm

Assuming a local pressure of 14.7 psi(a) and a temperature of 80°F (460 + 86 = 546 K) at a distance of 1 m away from the compressor inlet, the equivalent free air delivery can be calculated as follows:

$$\text{Free air delivery [cfm (fad)]} = 300 \times \frac{14.7}{14.5} \times \frac{460+86}{460+32}$$
$$= 338 \text{ cfm (fad)}$$

> Total length of straight pipes   = 200 ft
> Total length of pipes   = (200 + 100) = 300 ft [See the previous page]
> Pressure   = 100 psi = 114.5 psi(a)
> Pressure drop   = 14.5 psi

The internal pipe diameter (d)

$$d = \sqrt[5]{\frac{283 \times Q^{1.85} \times l}{P \times \Delta P}}$$

$$d = \sqrt[5]{\frac{283 \times 338^{1.85} \times 300}{114.5 \times 14.5}}$$

$$= 18 \text{ inch}$$

Chapter 6 | Secondary Air Treatment

The secondary air treatment is an effort to prepare compressed air finely, regulate pressure to the application requirement, and mix the air with a fine mist of lubricating oil just before the entry of compressed air into the associated application. The components used in the secondary air treatment are the filter, regulator, and lubricator (FRL).

Filter
A filter can be fitted to the branch line of a compressed air system to remove dust, dirt, and water from the compressed air.

The flow rate requirement of a filter in the branch line is much lower than that of the filter in the mainline.

Pressure regulator
Pneumatic machines and appliances require relatively steady pressure for their satisfactory operation. However, the operating pressure tends to fluctuate due to variations in the supply pressure or load pressure. It is, therefore, essential to regulate the operating pressure to match the requirements of the load, regardless of the variations in the supply pressure or the load pressure. Diaphragm regulators are the commonly used pressure regulators in industrial pneumatic systems.

Filter-regulator
This design combines the filter and regulator as a single unit. Air flows first through the filter and is then directed to the regulator. The advantage of this design is that only one unit is to be mounted, thus simplifying the installation work and reducing the cost.

Pressure gauge
The most commonly used device for measuring pressure is the bourdon tube.

Lubricator
The requirement of lubrication for the moving parts of components in a pneumatic system can be met by using valves and cylinders provided with an integral lubricant in each component or by injecting a controlled quantity of oil mist into the air stream using a mist lubricator.

Lubricators can be of the micro-fog type or the oil-fog type. The micro-fog type is used for the most general-purpose applications, and the oil-fog type is used for heavy-duty applications.

The lubricating oil is usually stored in a transparent polycarbonate bowl or a metal tank of large volume with a sight glass. A micro-fog lubricator cannot be filled under pressure except when fitted with a quick-fill device.

Air service unit
An FRL unit is shown in Figure 6.1. It comprises the following:
- Shut off valves to isolate upstream air and downstream air
- A combined filter and pressure regulator unit with a gauge
- Lubricator

For ease of use and system flexibility, a handy method of combining these units is to use a modular system. Typical specifications of pressure regulators and lubricators are given in Table 6.1 and Table 6.2.

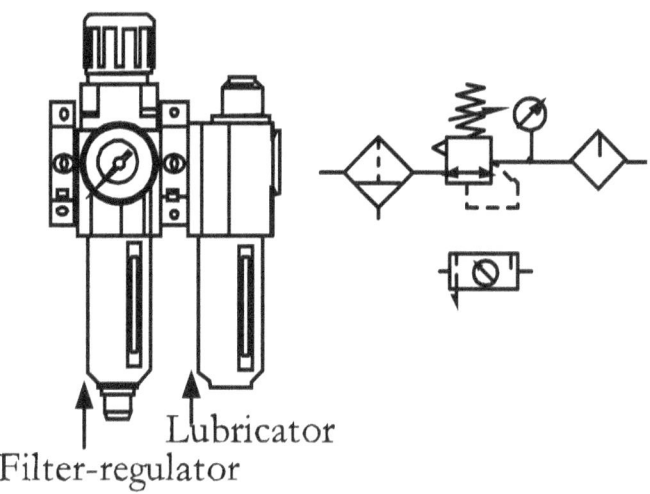

Figure 6.1 | Air service unit

Typical Specifications of Pressure Regulators

Table 6.1 | Specifications of pressure regulators

Flow scfm	Operating pressure range (psi)	Size
74	4.3 to 145	G¼
170	4.3 to 145	G⅜
254	4.3 to 145	G½
254	4.3 to 145	G¾
318	6 to 120	G¾
382	6 to 120	G1
382	6 to 120	G1¼
382	6 to 120	G1½

Regulator – with gauge / without gauge
Diaphragm: Relieving / Non-relieving
Threads: PTF / ISO Rc taper / ISO G parallel

Typical Specifications of Lubricators

Table 6.2 | Specifications of lubricators

Flow scfm	Bowl capacity (liter)	Operating pressure, psi	Size
53	12	0 – 230	G¼
131	12	0 – 230	G⅜
153	12	0 – 230	G½
153	12	0 – 230	G¾
223	30	0 – 230	G¾
297	30	0 – 230	G1
297	30	0 – 230	G1¼
297	30	0 – 230	G1½

Type: Oil-fog / Micro-fog
Bowl: Metal with level indicator / Transparent with guard
Threads: PTF / ISO Rc taper / ISO G parallel

[1 l/s = 2.12 scfm]

Chapter 7 | Pneumatic Actuators

There are two basic types of pneumatic actuators. They are: (1) Linear actuators and (2) Rotary actuators.

Linear Actuators

A pneumatic cylinder converts pneumatic power into a controllable linear force, motion, or both.

Terms and Definitions

Some essential parameters concerning pneumatic cylinders' operation and applications are their bore diameter, piston-rod diameter, force (thrust and pull), stroke length, speed, and piston-rod buckling.

Maximum Operating Pressure (P): It is the pressure that overcomes all resistances in the system, including useful work and losses. Alternatively, it is the maximum working pressure that the cylinder can sustain without adverse consequences.

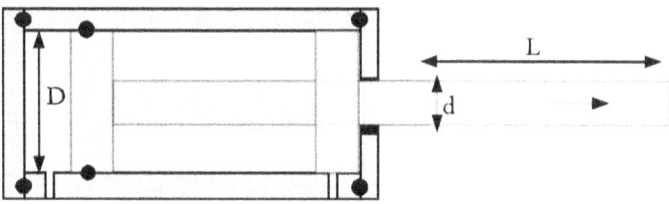

Figure 7.1 | Cylinder parameters

Bore Diameter (D): It refers to the diameter at the bore of the cylinder (Figure 7.1). It can be used to calculate the bore area of the cylinder. It is also equal to the piston diameter in a close-fitting pneumatic cylinder.

Piston-rod Diameter (d): It refers to the diameter of the piston-rod of the cylinder (Figure 7.1).

Maximum Stroke Length: It is the maximum linear movement that a cylinder can produce. The maximum stroke length for single-acting cylinders is typically 4 inches. The maximum stroke length for double-acting cylinders is 8 ft. For special designs, the stroke lengths can be up to 30 ft.

Thrust

The theoretical thrust (out-stroke) or pull (in-stroke) of a cylinder is the product of the effective piston area and the operating pressure. The effective area considered for the thrust calculation is the full area of the cylinder bore and is given by $\pi D^2/4$. The effective area for calculating pull is the full area of the cylinder bore minus the rod area and is given by $\pi (D^2-d^2)/4$.

$$\text{Thrust, F (lb)} = P \text{ (psi) x } A_{ext} \text{ (in}^2)$$
$$\text{Pull, F (lb)} = P \text{ (psi) x } A_{ret} \text{ (in}^2)$$

Where,

A_p is the piston area
A_r is the piston-rod area
A_{ext} is the active area during extension: ($A_{ext} = A_p$)
A_{ret} is the active area during retraction: ($A_{ret} = A_p - A_r$)

Table 7.1 gives the air consumption chart for pneumatic cylinders. Figures given in the tables do not make allowance for the loss due to packing friction or air leakage. This type of friction is estimated to affect the thrust of the cylinders by about 10%.

Since the air pressure in a plant may vary erratically due to intermittent use of large volumes of compressed air in all types of pneumatic equipment, the bore size of the cylinder must be large enough to provide the force required after allowing for any normal pressure drop.

Example 7.1
Determine the theoretical thrust and pull of a 2" bore double-acting pneumatic cylinder having a piston rod diameter of 0.75", supplied with compressed air at a pressure of 90 psi.

Solution
Bore diameter   = 2"
Piston-rod diameter   = 0.75"
Operating pressure   = 90 psi

$$\text{Thrust, F} = \frac{\pi \cdot 2^2}{4} \cdot 90 = 283 \text{ lb}$$

$$\text{Pull, F} = \frac{\pi \cdot \left(2^2 - 0.75^2\right)}{4} \cdot 90 = 243 \text{ lb}$$

Pneumatic Cylinder Forces

Thrusts of Single-acting Cylinders

Table 7.1 | Thrusts of single-acting cylinders

Piston diameter (in)	Effective area, in^2	Thrust in pounds @psi			
		80	90	100	120
5/16	0.08	6.4	7.2	8	9.6
7/16	0.15	12	13.5	15	18
9/16	0.25	20	22.5	25	30
¾	0.44	35.2	39.6	44	52.8
7/8	0.60	48	54	60	72
1-1/16	0.89	71.2	80.1	89	106.8
1¼	1.23	98.4	110.7	123	147.6
1½	1.77	141.6	159.3	177	212.4
1¾	2.41	192.8	216.9	241	289.2
2	3.14	251.2	282.6	314	376.8
2½	4.91	392,8	441.9	491	589.2
3.00	7.07	565.6	636.3	707	848.4

Thrusts and Pulls of Double-acting Cylinders

The values of cylinder thrusts in pounds against the piston area are given in Table 7.2.

Table 7.2 | Thrusts and pulls of double-acting cylinders

Piston diameter (D)	Rod diameter (d)	Effective area, in²	Thrust (Pull) in pounds @psi			
			80	90	100	120
5/16	0.125	0.06	6(5)	7(6)	8(6)	9(8)
9/16	0.187	0.22	20(18)	22(20)	(25)22	30(27)
1-1/16	0.25	0.84	71(67)	80(75)	89(84)	106(100)
1.50	5/8	1.46	141(117)	159(131)	177(146)	212(175)
1.50	1	0.98	141(79)	159(88)	177(98)	212(118)
2.00	5/8	2.83	251(227)	283(255)	314(283)	377(340)
2.00	1	2.36	251(188)	283(212)	314(236)	377(283)
2.50	5/8	4.60	368	414	460	552
2.50	1	4.12	330	371	412	495
3.00	5/8	6.76	565(541)	636(608)	707(676)	848)811
3.25	1	7.51	663(601)	746(676)	829(751)	995(901)
3.25	1⅜	6.81	663(545)	746(613)	829(681)	995(817)
4.00	1	11.78	1005(942)	1130(1060)	1256(1178)	1507(1413)
4.00	1⅜	11.08	1005(886)	1130(997)	1256(1108)	1507(1329)
5.00	1	18.84	1570(1507)	1766(1696)	1963(1884)	2355(2261)
5.00	1⅜	18.14	1570(1451)	1766(1633)	1963(1814)	2355(2177)
6	1⅜	26.78	2261(2142)	2543(2410)	2826(2678)	3391(3213)
6	1¾	25.86	2261(2068)	2543(2327)	2826(2586)	3391(3103)
8	1⅜	48.76	4019(3900)	4522(4388)	5024(4876)	6029(5851)
8	1¾	47.84	4019(3827)	4522(4305)	5024(4784)	6029(5740)
10	1¾	76.10	6280(6088)	7065(6849)	7850(7610)	9420(9132)
10	2	75.36	6280(6029)	7065(6782)	7850(7536)	9420(9043)
12	2	109.90	9043(8792)	10174(9891)	11304(10990)	13565(13188)
12	2½	108.13	9043(8651)	10174(9732)	11304(10813)	13565(12976)

Note:

[1 cu ft = 1728 cu.in]

Cylinder Air Consumption

The equations for the volume of free air displaced by the piston during the forward stroke and the in-stroke of a double-acting cylinder are given below:

$$V(\text{out-stroke}) = \frac{\pi D^2}{4} \, S \, \frac{Ps + Pa}{Pa}$$

$$V(\text{in-stroke}) = \frac{\pi (D^2 - \mathbf{d^2})}{4} \, S \, \frac{Ps + Pa}{Pa}$$

Where,

D=Cylinder bore, in
d=Rod diameter, in
V=Volume of free air, in^3
S=Stroke, in
P_s=Supply gauge pressure, psi
P_a=Atmospheric pressure (assumed to be 14.5 psi)
$(P_s + P_a)/Pa$=Compression ratio

The Compression ratio $(P_s + P_a)/P_a$ may be considered a multiplying factor to normalize the pressure condition.

Example 7.2

Calculate the air consumption per inch stroke of a double-acting cylinder with a 1½" bore and ⅜" piston-rod diameter supplied compressed air at a pressure of 90 psi.

Solution

Bore diameter = 1½ inch
Rod diameter   = ⅜ inch

$$V(\text{out-stroke}) = \frac{\pi \cdot 1.5^2}{4} \cdot 1 \cdot \frac{90 + 14.5}{14.5} = 12.73 \text{ in}^3/\text{inch}$$

$$=0.00737 \text{ cu ft (std)/inch stroke}$$

$$V(\text{in-stroke}) = \frac{\pi \cdot (1.5^2 - \mathbf{0.375^2})}{4} \cdot 1 \cdot \frac{90 + 14.5}{14.5} = 11.93 \text{ in}^3/\text{inch}$$

$$=0.0069 \text{ cu ft (std)/inch stroke}$$

$$\text{Volume (Total)} = 0.01427 \text{ cu ft (std)/inch stroke}$$

Air Consumption Chart for Pneumatic Cylinders

Table 7.3 gives the air consumption of pneumatic cylinders of different bore sizes for 1" stroke, full stroke. To get the air consumption for one full cycle, take the value corresponding to the bore size and pressure, and multiply it with the actual stroke.

Table 7.3 | Air consumption of pneumatic cylinders
(cu ft per inch of forward stroke)

Bore inch	Rod inch	Air consumption for the		
		forward stroke of 1 inch at 90 psi cu ft	return stroke of 1 inch at 90 psi cu ft	combined strokes of 1 inch at 90 psi cu ft
5/16	1/8	0.0003197	0.0002686	0.0005883
9/16	3/16	0.0010359	0.0009214	0.0019573
1 1/6	1/4	0.0044562	0.0042516	0.0087078
1 1/2	5/8	0.0073664	0.0060875	0.0134540
1 1/2	1	0.0073664	0.0040925	0.0114589
2	5/8	0.0130959	0.0118170	0.0249128
2	1	0.0130959	0.0098219	0.0229178
2 1/2	5/8	0.0204623	0.0191834	0.0396457
2 1/2	1	0.0204623	0.0171883	0.0376506
3	5/8	0.0294657	0.0281868	0.0576525
3 1/4	1	0.0345813	0.0313073	0.0658886
3 1/4	1 3/8	0.0345813	0.0283914	0.0629727
4	1	0.0523835	0.0491095	0.1014930
4	1 3/8	0.0523835	0.0461936	0.0985771
5	1	0.0818492	0.0785752	0.1604243
5	1 3/8	0.0818492	0.0756593	0.1575085
6	1 3/8	0.1178628	0.1116729	0.2295357
6	1 3/4	0.1178628	0.1078363	0.2256991
8	1 3/8	0.2095338	0.2033440	0.4128778
8	1 3/4	0.2095338	0.1995073	0.4090412
10	1 3/4	0.3273966	0.3173701	0.6447667
10	2	0.3273966	0.3143008	0.6416974
12	2	0.4714511	0.4583553	0.9298064
12	2 1/2	0.4714511	0.4509889	0.9224400

1. *Take each figure and multiply by the stroke in inches.*
2. *For pressures other than 90 psi, multiply the air consumption value by the given absolute pressure and divide it by 104.5 (90+14.5).*

[See Design Example 10.3 for more details]

Cylinder Speed

Assume that the piston-rod assembly of a cylinder moves with a velocity of 'v' when pushed by the system fluid with a flow rate of 'Q'. Further, assume that the cylinder piston of area 'A' has moved a distance 'S' in time 't' for attaining the velocity v.

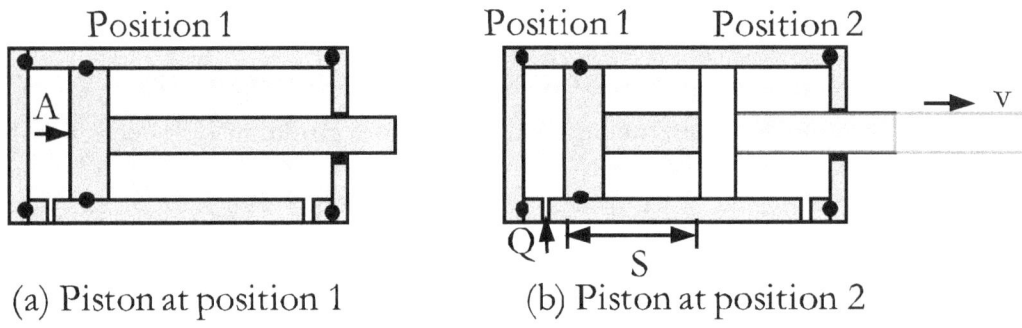

Figure 7.2 | Illustration of a cylinder in two piston positions

Figure 7.2 (a) and (b) show two positions (Position 1 and Position 2) of a cylinder for determining the speed during its forward stroke. Position 1 indicates the retracted position of the cylinder piston assembly, and Position 2 indicates an extended position. In Figure 7.2(b), both Position 1 and Position 2 are superimposed.

Mathematically,

$$v = S/t \qquad \text{or} \qquad t = S/v$$

We can easily relate the theoretical flow rate (Q) of the fluid to the speed (v) at which the piston-rod moves if we consider the cylinder volume (V) that must be filled with the fluid and the distance (S) through which the piston must travel at the specified speed. The volume (V) of the cylinder is the length of the stroke (S) multiplied by the piston area (A). The equation to determine the flow rate (Q) to achieve the required speed (v) is given below.

$$Q \ (\text{ft}^3/s) = \frac{V \ (\text{ft}^3)}{t \ (s)} = \frac{A \ (\text{ft}^2) \ x \ S \ (\text{ft})}{t \ (s)} = A \ (\text{ft}^2) \ x \ v \ (\text{ft}/s)$$

It can be observed from the equation mentioned above that the speed (v) of a given cylinder depends on the flow rate (Q) of the fluid.

Cylinder Mounting

A side load will be introduced if a cylinder is not adequately aligned with the load.

A pneumatic cylinder can be fixed using rigid mounting or pivot mounting. The rigid mounting methods include side, nose, flange, and face mounts. Next, the pivot mounting methods include the pivot, clevis, and trunnion. Installation requires a rod eye or rod clevis on the piston rod.

Cylinder Buckling

If a compressive axial load is to be applied to a long piston rod, it must be within the safety limit to prevent rod buckling. Due to buckling stress, the permissible load of the cylinder with a long stroke length is lesser than that ought to have been provided by the same maximum permissible working pressure and the piston surface area.

Limitations on Maximum Thrust Force

The piston-rod diameter and overall length limit the maximum thrust force which a cylinder can practically provide. The cylinder must also be supported adequately.

Note that a head-end mounting provides greater column strength than the cap-end mounting due to the smaller distance between the mounting points in the head-end mounting than that in the cap-end-mounting.

The piston-rod size of a pneumatic cylinder can be selected from the size charts with the help of values of its free buckling length or exposed length of the rod and the load imposed on the cylinder. The exposed length is typically longer than the stroke length of the cylinder. If the rod and the front end of the cylinder barrel are rigidly fixed, then the exposed length of the rod can be taken as half of the length of the actual piston rod. If pivot-to-pivot mounting is used, then the rod's exposed length can be twice the length of the actual piston rod.

Piston-rod Strength

Table 7.4 gives the suggested minimum piston rod diameters.

Table 7.4 | Suggested minimum piston rod diameters

Load	The exposed length of the piston rod in inches						
	10	20	40	60	80	100	120
2.5 lb		1/8					
11 lb	1/8	3/16					
14 lb	1/8	3/16					
56 lb	3/16	5/16					
400 lb	3/16	7/16					
667 lb	3/16	1/2					
1000 lb	1/4	1/2	3/4	1			
1500 lb	7/16	1/2	13/16	1-1/16			
1 ton	1/2	5/8	7/8	1-1/8	1-3/8		
1½ ton	1/2	11/16	15/16	1-3/16	1-1/2		
2 ton	1/2	3/4	1	1-1/4	1-9/16	1-13/16	
3 ton	13/16	7/8	1-1/8	1-3/8	1-5/8	1-7/8	
4 ton	15/16	1	1-3/16	1-1/2	1-3/4	2	2-1/4
5 ton	1	1-1/8	1-5/16	1-9/16	1-7/8	2-1/8	2-3/8
7½ ton	1-3/16	1-1/4	1-7/16	1-3/4	2	2-1/4	2-1/2
10 ton	1-3/8	1-7/16	1-5/8	1-7/8	2-1/8	2-7/16	2-3/4
15 ton	1-11/16	1-3/4	1-7/8	2-1/8	2-3/8	2-11/16	3
20 ton	2	2	2-1/8	2-3/8	2-5/8	2-7/8	3-1/4
30 ton	2-3/8	2-7/16	2-1/2	2-3/4	2-7/8	3-1/4	3-1/2
40 ton	2-3/4	2-3/4	2-7/8	3	3-1/4	3-1/2	3-3/4
50 ton	3-1/8	3-1/8	3-1/4	3-3/8	3-1/2	3-3/4	4
75 ton	3-3/4	3-3/4	3-7/8	4	4-1/8	4-3/8	4-1/2
100 ton	4-3/8	4-3/8	4-3/8	4-1/2	4-3/4	4-7/8	5
150 ton	5-3/8	5-3/8	5-3/8	5-1/2	5-1/2	5-3/4	6

[1 us ton = 2000 pound]

Technical Data for Pneumatic Cylinders

Standard values of cylinder diameters and maximum stroke lengths are presented in Table 7.5.

Table 7.5 | Standard values

Parameters	Standard values
Bore diameter (in)	5/16, 9/16, ¾, 1, 1-1/16, 1⅛, 1½, 2, 2½, 3, 3¼, 4, 5, 6, 8, 10, 12
Rod diameter (in)	0.03 to 2.5
Stroke (ft)	0.03 to 6.5
Operating pressure (psi)	Up to 175

Materials Used for Pneumatic Cylinders

Table 7.6 gives the materials used for the construction of pneumatic cylinders.

Table 7.6 | Materials used for cylinders

Material	Trade Name	Part
Brass. Steel, X 5 Cr Ni 18 9	-	Barrel
X 20 Cr 13	-	Piston rods
Cast iron	-	Mounting parts
Nitrile Buna Rubber (NBR)	Perbunan	Seals
Fluroelastomeric rubber (FKM)	Viton	Heat-resistant seals
Polyurethane	-	Diaphragm

Standards for Pneumatic Cylinders

Table 7.7 gives the standards for pneumatic cylinders.

Table 7.7 | Pneumatic cylinder standards

Pneumatic component	Standard	Remarks
Standard cylinder	ISO 6431	International
	ISO 6432	International
	NFPA (JIC)	USA
	VDMA 24652	German
	NFE 49003.1	France

Rotary Actuators

A rotary actuator converts the energy of the compressed air into rotary mechanical energy. Air motors are designed for continuous rotation. Semi-rotary actuators are designed for reciprocating rotary motion.

A rotary actuator can be defined by its torque and running speed. The starting torque of a rotary actuator is the torque available to move a connected load from rest. Stall torque is the torque the load must apply to bring a running actuator to rest. Running torque is the torque available at any given speed.

An air motor can withstand repeated stalling and reversing without harm or overheating. It can accelerate rapidly since the energy in the compressed air is released at a high rate.

Semi-rotary Actuators

Semi-rotary actuators are constructed with a rotating vane or with a rack-and-pinion design. The vane type rotary actuator with limited travel, as shown in Figure 7.3, consists of a single vane coupled to the output shaft. It is usually designed for a double-acting operation with a maximum angle of rotation of 270°. Usually, the angle of rotation can be adjusted.

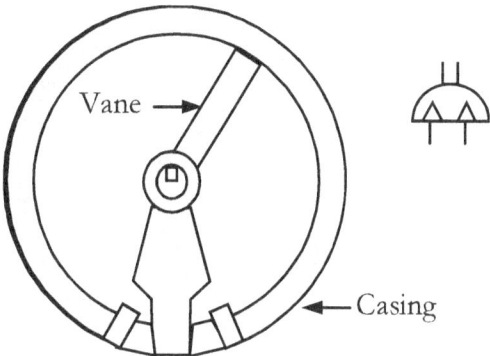

Figure 7.3 | Semi-rotary actuator – Vane type

The rack-and-pinion type of rotary actuator with limited travel is shown in Figure 7.4. It consists of a double-acting piston coupled to the output shaft by a rack-and-pinion arrangement. An angle of rotation up to 360° is possible with this type of design.

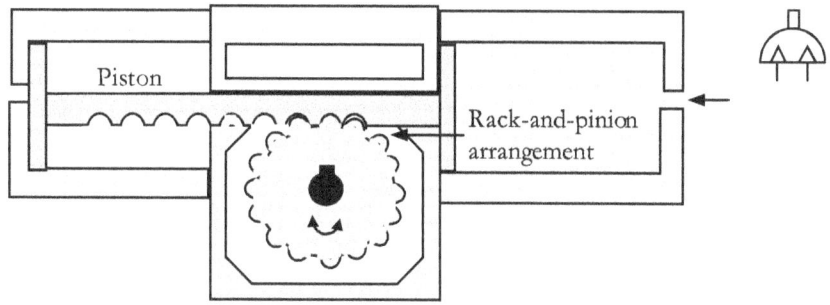

Figure 7.4 | Semi-rotary actuator – Rack-and-pinion type

Air Motors

Air motors convert the potential energy of compressed air into rotary mechanical energy. They are designed to provide continuous rotation. Piston, vane, and gear designs are generally used for air motors.

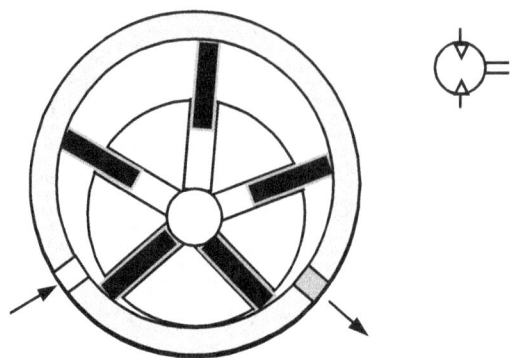

Figure 7.5 | Sliding vane motors

A rotary vane motor is shown in Figure 7.5. It consists of a cylindrical rotor with sliding vanes placed eccentrically in a cylindrical housing. As air enters the inlet port and passes into the cylinder, a pressure unbalance acts on the vanes. This pressure unbalance develops a torque that turns the rotor against the motor's load.

Piston motors have four to six cylinders arranged in radial or axial positions. The cylinder pistons reciprocate in sequence when compressed air is applied. These piston movements cause a crankshaft to turn through a connecting rod, thus causing the rotation of the output shaft.

In the gear design, torque is generated by the teeth profiles of two meshed gear wheels.

Typically, air motors have power ratings ranging from ⅛ to 25 hp. Many speed ranges are possible, ranging from as low as 40 rpm up to as high as 50000 rpm.

Technical Data for Rotary Actuators

Standard values of sizes and rotating angles of rotary actuators are presented in Table 7.8.

Table 7.8 | Standard values

Parameters	Standard values	Remarks
Standard sizes of rotary actuators	0.2 to 4 in	-
Standard rotating angle for rotary actuators	90^0, 180^0, 270^0, 360^0	Fixed
	30^0 to 270^0	Variable
Torque	1.3 to 1300 lb.in	
Operating pressure	Up to 175 psi	

Note: Additional data for different types of actuators are given in Appendix 2

Pneumatic Tools

Compressed air permits the use of tools that are compact, light in weight, portable, and easy to operate. Pneumatic portable tools perform various operations such as nut running, screw driving, grinding, drilling, riveting, scaling, stud driving, and wire wrapping. Portable tools include screwdrivers, hammers, riveters, abrasive tools, and hoists. Table 7.9 gives the air consumption chart for industrial-type tools.

Air Consumption Chart for Industrial Type Tools

Table 7.9 | Air consumption at 70 to 90 psi

Tool	Consumption (cfm) at 25% usage factor
Air motor, 1 hp to 3 hp	9 to 24
Burring tool, small to large	4 to 6
Chipping hammer	8
Die grinder, medium	6
Drill, 1/16" to 5/8"	6 to 9
Horizontal grinder, 2", 4", 6", 8"	8 to 20
Impact Wrench, 1/4" to 1¼"	4 to 14
Nut setters, large up to 3/8" to 3/4"	6 to 15
Paint spray gun	5
Rammers (small, medium, and large)	6 to 10
Riveting hammer (light and heavy)	4 to 8
Saws, circular	16
Scaling hammer	3
Screwdriver	3 to 6
Trapper, up to 3/8"	6
Vertical grinders and sanders	9 to 20

1. *Air consumption is only indicative and may not be accurate for any particular make*
2. *Always check with the OEM for the actual air consumption of tools*

Note: Additional data for pneumatic tools are given in Appendix 3

Vacuum Grippers

A vacuum gripper comprises a vacuum generator (ejector) and a suction cup. An air suction or vacuum filter is provided in the vacuum passage to prevent dust intrusion into the ejector. Figure 7.6 shows a vacuum generator and suction cups.

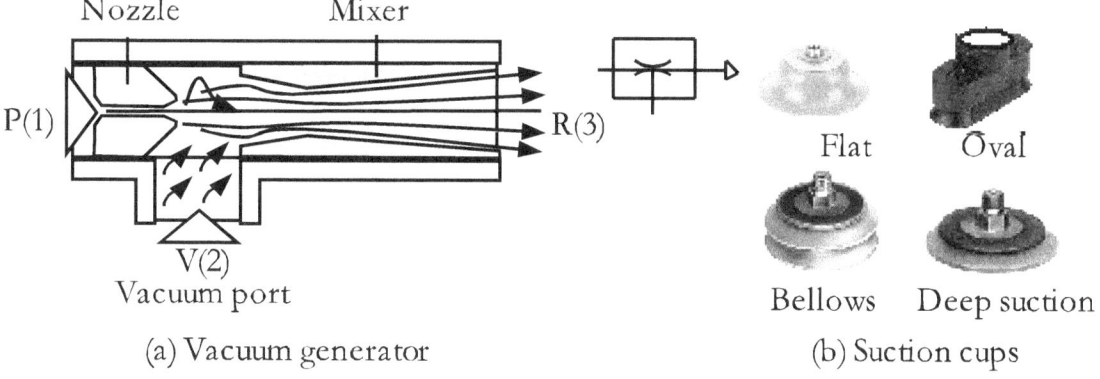

(a) Vacuum generator   (b) Suction cups

Figure 7.6 | Vacuum generator and Suction cups

61

Vacuum Gripper Terms

Maximum Suction Flow Rate: It is the maximum value of the volume of air taken in without having anything connected to the vacuum port.

Maximum Vacuum Pressure: It is the maximum value of the vacuum pressure generated by the ejector.

Air Consumption: It is the volume of the compressed air consumed by the ejector.

Typical Specifications, Vacuum Grippers

Table 7.10 gives typical specifications of vacuum generators.

Table 7.10 | Typical specifications of vacuum generators

Nozzle diameter (in)	Max. suction flow (scfm)	Air consumption (scfm)
0.02	0.22	0.47
0.04	0.94	1.87
0.05	1.44	3.02
0.06	2.09	4.07
0.07	2.74	5.83
0.08	3.24	196

Suction Cups

Suction cups are used to transport workpieces of different weights, surfaces, and shapes.

They are made from nitrile rubber, polyurethane, and silicone for use in various applications. Suction cups made from silicone are food-safe.

Round flat type suction cups can retain workpieces with smooth, impervious surfaces. However, round bellows-type suction cups can adapt to suit uneven, curved, and inclined surfaces.

Typical nominal diameters in inches for suction cups are as follows: 0.1, 0.2, 0.3, 0.4, 0.6, 1.2, 1.6, 2.2, 3.0, 4.0, and 5.0.

Typical Specifications of Round Flat Suction Cups

Table 7.11 gives typical specifications of suction cups.

Table 7.11 | Typical specifications of round flat suction cups

Suction cup diameter, inch	Holding force (lb)	Vacuum connection
0.1	0.03	M3
0.2	0.20	M5
0.3	0.36	M5
0.4	1.01	M5
0.6	1.78	$G\frac{1}{8}$
1.2	7.64	$G\frac{1}{8}$
1.6	12.59	$G\frac{1}{4}$
2.2	23.83	$G\frac{1}{4}$
3.0	44.29	$G\frac{1}{4}$
4.0	89.25	$G\frac{1}{4}$
5.0	136.23	$G\frac{3}{8}$

Holding force at nominal operating pressure -0.7 bar

Chapter 8 | Pneumatic Valves

A pneumatic valve consists of a body with an internal moving element, such as a poppet or spool, actuating mechanisms, and many ports. It is a control device that directs or restricts the compressed air flow or controls the flow based on a specified pressure condition in a particular part of the associated circuit. Accordingly, pneumatic valves can be classified as directional control valves, flow control valves, and pressure control valves.

Directional Control Valves
A directional Control (DC) valve (or way valve) controls the path the compressed air takes. A non-return valve (NRV) is a directional control valve that allows the compressed air flow in a pneumatic system in only one direction and blocks the flow in the opposite direction.

3/2-DC Valves (NC Type)
3/2-way normally-closed valves are used as final control elements to control single-acting cylinders, unidirectional motors, and other valves. The symbol of a 3/2 DC valve is given in Figure 8.1.

5/2-Directional Control Valves
5/2-DC valves are used as final control elements to control double-acting cylinders. The symbol of a 5/2 DC valve is given in Figure 8.1.

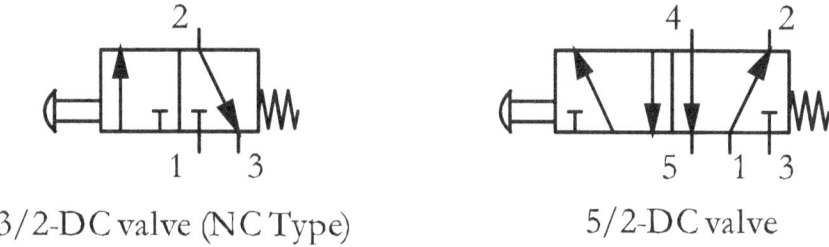

3/2-DC valve (NC Type)   5/2-DC valve

Figure 8.1 | Symbols of directional control valves

Check Valves and Flow Control Valves
A flow control valve restricts the flow rate at which pressurized fluid is transferred in a pneumatic circuit. The symbols of a check valve and a one-way flow control valve are given in Figure 8.2.

Check valve   One-way flow control valve

Figure 8.2 | Symbols of a check valve and a flow control valve

Pressure Control Valve
A pressure sequence valve generates a control signal when a set pressure in a particular part of the system has been reached for initiating a subsequent action.

Pneumatic Valve Sizing

Many methods aid in selecting pneumatic valves used as final control elements. Two main methods of sizing pneumatic valves are: (1) finding the value of the coefficient of flow Cv (or Kv) parameter or (2) using a sizing table that relates Cv values to the cylinder bore sizes.

The value of Cv (valve coefficient) required to operate a given cylinder for a specific time is given below.

$$C_v = \frac{\text{Area x Stroke x A x } C_f}{\text{Time x 29}}$$

Where,
Cylinder area in square inch
Cylinder stroke in inch
A = Pressure drop constant (From Table 8.1)
C_f = Compression factor (From Table 8.1)
Time in seconds

The compression factor C_f and the pressure drop constant A can be found in Table 8.1.

Table 8.1 | Compression factor C_f and pressure drop constant

Inlet pressure psi	Compression factor, C_f	Constant A at the pressure drop		
		2 psi	5 psi	10 psi
10	1.6	--	0.102	--
20	2.3	0.129	0.083	0.066
30	3.0	0.113	0.072	0.055
40	3.7	0.097	0.064	0.048
50	4.4	0.091	0.059	0.043
60	5.1	0.084	0.054	0.040
70	5.7	0.079	0.050	0.037
80	6.4	0.075	0.048	0.035
90	7.1	0.071	0.045	0.033
100	7.8	0.068	0.043	0.031
110	8.5	0.065	0.041	0.030
120	9.2	0.062	0.039	0.29

For most applications, the constant A is selected for a pressure drop of 5 psi. The higher the Cv, the greater the flow. To account for losses, oversize a valve by at least 25%.

The sizing chart in Table 8.2 indexes the Cv values to the cylinder bore sizes (in) and the cylinder's operating speeds (inches/s).

Table 8.2 | Sizing chart for pneumatic valves [Speed (inches/s)]

Cv	Bore size (in)									
	0.75	1.13	1.5	2.0	2.5	3.25	4.0	5.0	6.0	8.0
0.1	26.8	11.9	6.7	3.8	2.4	1.4	0.94	0.6	0.42	0.24
0.2	53.7	23.9	13.4	7.5	4.8	2.9	1.9	1.2	0.84	0.47
0.5	134	59.6	33.6	18.9	12.1	7.1	4.7	3	2.1	1.2
1.0	268	119	67.1	37.7	24.2	14.3	9.4	6	4.2	2.4
2.0	537	239	134	75.5	48.3	28.6	18.9	12.1	8.4	4.7
4.0		477	268	151	96.6	57.2	37.7	24.2	16.8	9.4
8.0			536	302	193	114	75.5	48.3	33.6	18.9
16				604	387	229	151	96.6	67.1	37.7
32					773	457	302	193	134	75.5

Assumption: P = 80 psi, ΔP = 80%

The selection of a pneumatic valve depends on the associated cylinder's speed requirement. A pneumatic valve is designed with the nominal flow rate as an essential parameter. This parameter must be sufficient to meet the air consumption requirement of the cylinder for its satisfactory operation. It may be noted that the port size of the valve should match the port size of the associated cylinder. Another critical parameter for the valve selection is the permissible pressure drop across the valve.

The empirical values for the nominal flow rate provided by the manufacturers can be used as a guide for dimensioning pneumatic valves. Typical values of nominal flow rates are given in Table 8.3. Using these figures makes it possible to achieve sufficient cylinder speeds in most practical cases.

Table 8.3 | Standard nominal flow rate for valves

Cylinder piston dia (in)	Valve connection size	Approx. Nominal size (in)	Approx. Standard nominal flow rate (scfm)
Up to 1	M5	0.1	3
1 – 2	G⅛	0.15	Up to 6
2 – 4	G¼	0.3	Up to 40
4 – 8	G½	0.5	Up to 100
8 – 12	G¾/G1	0.75	Up to 200

Its desired function mainly decides the selection of other pneumatic valves used in signal processing in a control circuit. For most applications using trip cam and pilot valves, G⅛ poppet models are suitable. Where space is at a premium, and the valve being controlled is relatively small, M5 models will suffice.

Chapter 9 | Energy Efficiency of Compressed Air Systems

Careful planning is essential for designing an efficient pneumatic system. The overall cost of compressed air consists of capital costs and operating costs. The capital cost includes the cost of the compressor, air preparation units, valves, actuators, and conductors. The operating cost includes the cost of electricity, maintenance, and leakages. Electricity is needed for running the main drive for compressing air, cooling fans, etc. Remember, compressors are generally the biggest electricity consumers in compressed air systems. In the long run, the cumulative cost of electricity can go very high. Therefore, it is essential to size properly, lay the components and conductor system, and operate the system's critical components at their maximum efficiency to reduce losses and obtain the highest efficiency. Also, remember that an air compressor operates most efficiently at its full load.

Parameters Influencing the Efficiency

The starting point in designing a compressed air system is to accurately evaluate system actuators' maximum air consumption requirement at any given period. Remember, the air consumption requirement can be constant or varying depending upon the application. A compressor system must cost-effectively provide sufficient compressed air delivery at the required quantity, quality, and pressure to satisfy the compressed air demand of all air-consuming devices for smoothening the energy-intensive partial phases of load requirements. Further, the compressed air system is cost-effective if it fits its location and operating conditions.

The important design parameters that influence the efficiency of a pneumatic system are the delivery of the associated compressor, the maximum pressure setting in the system, the pressure drops across the components and pipes, the duty cycle, leakage, and the layout of its distribution system. Deciding the required air quality levels is important to determine the filtration and drying system types, component efficiencies, and affordable component cost. Also, remember that individual applications requiring high-quality air should not dictate the design of the overall system. Apart from the design parameters, how the system is maintained also affects its efficiency during its service life.

A single-compressor or multiple-compressor unit can supply a plant's total compressed air requirement. A single-compressor unit is best suited for a small plant with constant compressed air demand. Depending on the usage profile, it can be more cost-effective to incorporate multiple compressors to take care of load variations. A multiple-compressor unit is well suited for a large system with wide fluctuations in compressed air demand within a shift of operations or among the shifts. It is also suited for a system that requires maintenance flexibility and a backup compressor unit. In some cases, compressors driven by variable frequency drives offer the best solution for controlling partial loads. Further, sequencers can improve the efficiency and reliability of multiple compressor systems. With microprocessor controls, they can stabilize system pressure and track the load.

An appropriately-sized reservoir can meet occasional peak demand from the system with a small compressor and avoid high electrical demand charges.

Operate a pneumatic system at the lowest possible pressure.

Compressed Air Demand

In a pneumatic system, cylinders, air-operated tools, etc., consume compressed air. To ensure efficient usage, it's crucial to determine the specific compressed air requirement for each consumer. The air consumption can be calculated using relevant mathematical formulas or obtained from the manufacturer's catalogs. The total compressed air demand for the system is then calculated by adding up the air consumption requirements of all individual air consumers at a particular time in the cycle of operation.

Variations in Compressed Air Demand

It's common for a compressed air system in a production unit to undergo significant fluctuations in demand over the course of a working day due to the nature of work operations. Additionally, large temperature swings can also cause demand fluctuations.

Classification of Air Demand

Compressors can be classified based on air demand as base load, peak load, or standby units.

Base Load Compressor

The term 'base load air demand' refers to the constant volume of compressed air required by a production facility. A base load compressor is responsible for providing the base load air demand.

Peak Load Compressors

On the other hand, the term 'peak load air demand' refers to the amount of compressed air needed during specific times of high demand. This volume fluctuates due to varying demands from different consumers. To fulfill peak load demand in a system, additional compressors are required.

Standby Compressor

It's standard practice in a compressed air station to have a backup compressor available when it needs to be taken offline for maintenance or replacement. This standby compressor can also be utilized to manage unexpected spikes in demand.

Compressed Air Delivery

Compressed air production must be precisely matched to suit constant or fluctuating compressed air demand. A compressed air system consists of a single compressor or multiple compressors of similar or varying sizes to meet a constant or varying compressed air demand. A single compressor installation can be selected for a smaller system, or a system constantly operates at full load output. A varying demand for compressed air in a large-scale production operation can be met using a correctly-sized large compressor or correctly graded multiple compressors of equal or differing capacities. An example of grading compressors is given below.

The design focus is the optimum energy efficiency of compressed air production. The requirement of optimum efficiency can be achieved by selecting compressors of appropriate sizes according to the base load and peak load compressed air demand of a production system. Coordinating the compressor operation is essential to ensure optimum energy efficiency using an appropriate control system.

Control System, Compressed Air Production

A control system (controller) is required to maintain stable system pressure for a compressed air system. In a multiple compressor system, the controller ensures that only needed compressors are active at their most efficient level. Therefore, a controller must be designed to precisely match the delivery of a compressor to the compressed air demand, thereby coordinating and optimizing the interplay between compressed air generation and consumption.

The operation of a compressor needs to be controlled to adjust the pressure and flow as per the demands of the application. Many control schemes are devised to suit different requirements, such as constant speed, variable speed, or intermittent drive operation. The control scheme includes the load/unload method, inlet valve modulation, and motor start-stop control.

The load/unload method allows a compressor to run at full load or no load while its drive motor remains at a constant speed. The inlet valve modulation throttles the compressor's inlet to change the compressor's delivery to meet the varying demand for compressed air. The start/stop control starts or stops the drive motor of a compressor to store compressed air in the associated receiver tank and supply a sufficient quantity of compressed air to meet the application demand.

A constant-speed control can be employed for a compressor in a steady pneumatic system when the consumption of compressed air is more than 75% of the compressor delivery, or the start-stop frequency of the associated motor exceeds the manufacturer's recommendations. The load/unload method or inlet valve modulation can be used for constant-speed control.

The start-stop (on-off) control can be used for a compressor when the compressed air consumption is less than 75% of the compressor delivery, and an adequate compressed air storage facility is provided. This type of control can be used for applications where compressed air is not required continually, allowing the compressor to get sufficient cooling time.

Rotary compressors (screw or vane compressors) are compressor stations' most heavily loaded equipment. They must be designed to provide maximum efficiency without running at extreme speeds to save energy, maximize service life, and enhance reliability. Therefore, a variable-speed drive can be employed in a screw compressor or vane compressor when energy efficiency is an important criterion. This type of control will dynamically adjust the drive's speed to meet the varying demands of compressed air in an application.

A dual control system can be used for a compressor whose operation varies between intermittent and continuous duty according to compressed air usage. A selector switch can realize the switching between constant-speed and start-stop operations. Next, sequencing controls can be employed for a group of compressors to operate at peak efficiency levels.

A PLC-based controller in a compressed air system with multiple compressors can shut off compressors not needed and bring the backup compressor when required to match compressor supply to demand.

A modern controller can sequence compressors as required in a compressed air system with multiple compressors. It can also ensure that fewer compressors operate at a lower pressure when compared to a system with a conventional controller.

A perfectly matched control system can help by increasing the load factor to over 90 % and achieving power savings of up to 20 %.

Master Controllers

A microprocessor-based controller can maintain stable system pressure and ensure that only necessary compressor units are operated optimally in a pneumatic system that uses multiple compressors. With a PLC-based controller, the compressor delivery can efficiently match demand. Compressors that are not required can be automatically shut off, while backup units can be brought in when necessary. Furthermore, advanced controllers can sequence and select compressors to further optimize the system's performance.

Chapter 10 | Design of Pneumatic Systems

A pneumatic system includes many critical components, such as a compressor, actuators, and control valves, interconnected using pipes and tubes. The energy transfer takes place through the medium of compressed air.

The design of a pneumatic system involves the determination of the force and speed requirements of all actuators, the selection and sizing of components, the air consumption rate of actuators, finding the sequence of operation, the determination of the required pressure level, and finding many other component parameters to meet the design objectives.

The optimum design of a pneumatic system for a project must try to synchronize with the availability and quality of components in the market.

Critical Design Steps

Building the right pneumatic system for the specific application requirements is best achieved by first determining the parameters of the components of the system. That is, the design process of a pneumatic system primarily consists of finding the sizes/capacities of various system components, such as cylinders, rotary actuators, valves, compressor(s), drive motor, tank, coolers, filters, dryers, and fluid conductors. The following critical steps may be followed to find the significant parameters while designing a pneumatic system with a compressor and cylinder.

An analysis of the system to be designed would reveal the application requirements of output force/torque, speed, and mechanical power output. By applying sound engineering principles, it is also required to identify the nature of each load to be moved in the system and establish the useful load to be handled by each actuator. The effects of friction, gravity, mass, inertia, and other external forces likely to be encountered in the system must be considered. The power transmission mechanism between each load and the corresponding actuator output shaft must be considered while calculating the useful load.

Select the operating pressure for a pneumatic system with a standard pressure range making allowances for the potential pressure losses and friction in the system. The most economical pressure for industrial pneumatic systems is 90 psi. Most actuators and air tools require 90 to 100 psi. However, many other actuators and tools may require higher pressures.

Determine the size of the cylinder by taking into account the force, pressure, speed, and stroke requirements of the cylinder and the nature of the load. The dimensions of the cylinder should be such that the force developed by it must overcome the load and frictional forces. The cylinder must also provide adequate acceleration. A single-acting cylinder can be used where the work operation is to be carried out only in one direction of motion of the cylinder. A double-acting cylinder should be used where a positive pneumatic return action is required and the cycle time is critical. Determine the cylinder's area and diameter (bore size) using the formulae on pages 51 and 52.

Also, decide the diameter of the cylinder piston-rod. A large piston-rod gives a large column strength and compression strength for the piston-rod.

Determine the required stroke length of the cylinder based on the data derived from the system analysis.

Cylinder extension and retraction speeds can be decided as per the system requirements.

Determine the cylinder's air consumption rate to meet an application's speed requirements. To estimate the total average air consumption of a pneumatic system, calculate the air consumption for each cylinder using the formulae given on page 54. Values of air consumption for forward and return strokes of pneumatic cylinders are also given in Table 7.3 (Page 55). Add the estimated air consumption of all cylinders and add 5% to make allowance for the loss due to leakage and friction.

Determine the air consumption rate of various tools, vacuum generators, and air motors to meet the speed requirements of the application.

Once the air requirement of each air-consuming device has been determined, calculate the total air requirement of all the devices for intermittent or continuous operations.

Find the duty cycle of the application. It attempts to find the number of hours the compressor is expected to run. For example, if 70% of the time the compressor is expected to provide air to the equipment, the amount of air to be supplied would be 0.7 times the total air requirement, as calculated in step 2.

Add an extra 30% to provide a reasonable buffer. Also, consider the requirements for future expansion.

However, it may be noted that the total air consumption, as calculated above, may give rise to excessive capacity requirements in the system. Therefore, one may have to be careful in determining the maximum flow rate expected under the actual service conditions.

Next, select a compressor and drive motor with a sufficient delivery rate to meet the peak air consumption rate of all actuators. Draw the important compressor specification parameters such as the drive motor power, drive speed, pressure rating, number of compression stages, operating voltage, etc. Remember, the rule of thumb is that on a steady pumping, a compressor will produce a minimum of 4 scfm flow of air for every HP capacity at 90 psi.

Select the size of the air receiver tank from the sizes offered by the compressor manufacturer.

Select the specifications of components, such as the aftercooler, mainline filter, dryer(s), and FRL unit(s), based on the flow rate.

Design the mainline and distribution conductor system to keep the pressure drop across the conductor system to a permissible limit. The selection of pipe size is governed by the delivery volume, required pipe length, operating pressure, and the permissible pressure drop.

Next, size all final control elements to meet the air consumption requirement of the actuators, using charts/data provided by manufacturers.

If the velocity of the piston-and-piston-rod assembly is high enough to cause end-of-stroke impact shocks or if there is no speed control feature incorporated for the cylinder, cushioning devices are essential for the cylinder. Typically cushioning is required if the velocity of the cylinder is greater than 0.3 ft/s.

The cylinder's permissible piston-rod buckling force is an important parameter to consider.
.
Next, five sample design problems are presented to highlight the essential design steps. However, it may be remembered that the optimum design depends on the site conditions and exact data available in the manufacturer's domain.

A simplified layout of a pneumatic system is given in Figure 10.1. a summary of typical design steps for a pneumatic system is given in Figure 10.2.

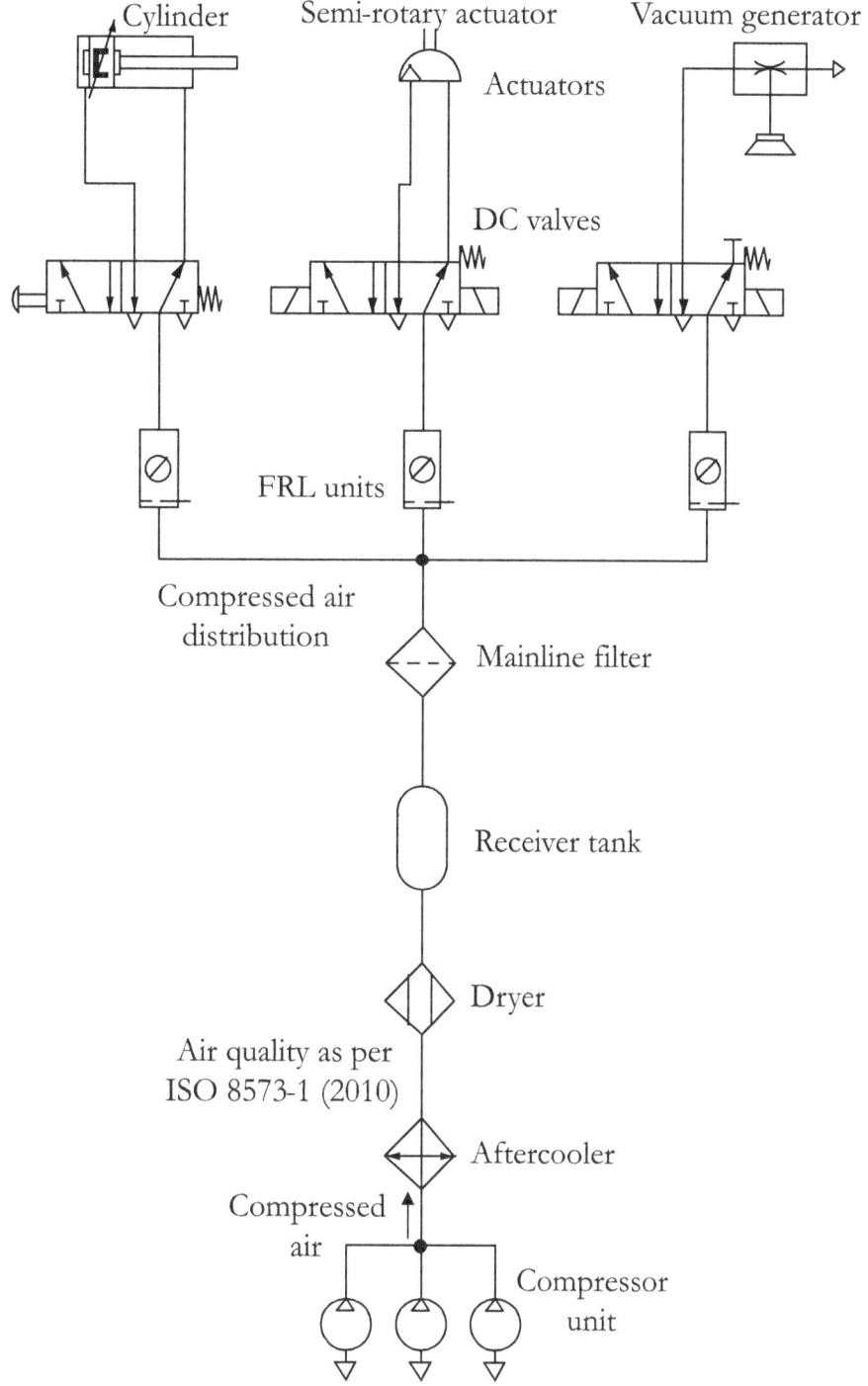

Figure 10.1 | A typical pneumatic system (Simplified)

A Summary of Typical Design Steps for a Pneumatic System
System analysis to understand work operations, their sequence, air quality, etc
Determine the types and number of actuators
Determine the requirements of force and speed of each actuator
Determine specifications of all actuators (Bore diameter, stroke length, etc)
Determine the air consumption demand of each actuator
Determine the total air consumption demand of all actuators
Determine the operating pressure based on the pressure ratings of actuators
Determine the compressor configuration based on the compressed air demand
Determine the specifications (capacity and pressure rating) of compressors
Design the control scheme for the compressors
Determine the specifications (type, size, accessories) of the air receiver tank
Determine the specifications (type, size, approach temp) of the aftercooler
Determine the specifications of dryers based on the air quality requirement
Determine the specifications of filters based on the air quality requirement
Determine the layout and size of distribution pipes, fittings, etc
Develop and simulate control circuits to realize the work operations
Determine the specifications (size, port, mounting) of control components
Procure components based on the specifications
Develop prototype and evaluate performance

Figure 10.2 | A summary of typical design steps for a pneumatic system

Design Problem 10.1

A service station uses the following types of tools and equipment with the average compressed air requirement, as given in Table 9.1. Determine the size of the compressor, the rating of the motor, and the tank size.

Table 10.1 | Data and calculations for the design problem 10.1

Tool	Qty	Air consumption per tool (cfm)	Total air consumption [cfm (FAD)]	Max. pressure (psi)
Blow Gun	1	2.5	2.5	90
Car Lift, 8000 lb	2	6.0	12.0	150
Grease Gun	2	3.0	6.0	125
Spark Plug Cleaner	1	5.0	5.0	90
Spring Oiler	1	4.0	4.0	90
Tire Inflation Line	1	2.0	2.0	125
Transmission Flusher	1	3.0	3.0	90
		Total	34.5	

Solution

Total air consumption = 34.5 cfm (FAD)

Operating pressure = 90 psi

Pressure, lower setting = 100 psi

Pressure, upper setting = 115 psi

Select the compressor specifications from the manufacturer's chart (See Table A1.3, Page 89)

- Compressor type = Rotary screw

- No. of stages = One

- Power, compressor = 10 hp

- Voltage = 460 V, 3 phase

- Maximum pressure = 150 psi

- Tank size = 80 gallons

Design Problem 10.2

A service station uses the following tools and equipment with the continuous compressed air requirement, as given in Table 9.2. Determine the size of the compressor, the rating of the motor, and the tank size.

Table 10.2 | Data and calculations for the design problem 10.2

Tool	Quantity	Air consumption per tool (cfm)	Total air consumption [cfm (FAD)]	Max. pressure (psi)
Air hammer	1	4	4	100
Body polisher	1	20	20	100
Paint spray gun (Production)	1	8	8	100
Touch-up spray gun	1	3.5	3.5	100
Vacuum cleaner	1	7	7	150
		Total	42.5	

Solution

Total air consumption = 42.5 cfm (Free air)

Operating pressure = 100 psi

Pressure, lower setting = 115 psi

Pressure, upper setting = 130 psi

Select the compressor specifications from the manufacturer's chart (See Table 3.4, Page 21)

- Compressor type = Reciprocating

- No. of stages = Single

- Power, compressor = 10 hp

- Voltage = 460 V, 3 phase

- Maximum pressure = 125 psi

- Tank size = 110 gallons

74

Design Problem 10.3

A double-acting cylinder is used for the clamping operation of a casting during machining. The cylinder has a bore diameter of 2½ inches, a rod diameter of 1 inch, and a stroke length of 8 inches. On average, the cylinder is to clamp 16 castings every minute. The system pressure is set at 100 psi. Find the air consumption of the cylinder at 100 psi.

Solution

Bore diameter	= 2½ inches
Rod diameter	= 1 inch
Stroke length	= 8 inches
No. of operations/minute	= 16

Referring to the air consumption chart Table 7.3, page 51,

Air consumption during the forward and return strokes at 90 psi

$$= 0.0376506 \ \text{ft}^3/\text{in of stroke}$$

$$= 0.0376506 \times 8 \ \text{ft}^3/\text{cycle}$$

$$= 0.0376506 \times 8 \times 16 \ \text{cfm(fad)}$$

$$= 4.8192768 \ \text{cfm(fad)}$$

Air consumption during the forward and return strokes at 100 psi

$$= 4.8192768 \times [(100+14.5)/(90+14.5)]$$

$$= 5.28 \ \text{cfm(fad)}$$

Note:

Calculate each cylinder's air consumption to estimate the average air consumption of a typical pneumatic system. Add the estimated air consumption of all cylinders and add 5% to make allowance for the loss due to leakage and friction.

1 cu ft = 1728 cu.in

Design problem 10.4
A manufacturing plant has 10 Nos. of automatic production machines using the following pneumatic actuators. The operating pressure is set at 90 psi. The duty cycle can be taken as 70%. Determine the size of the compressor, the rating of the motor, and the tank size. Assume, for each linear actuator, 8 operations (average) per minute.

Table 10.3 | Data and calculation for the design problem 10.4

Qty	Air consumption per inch stroke per actuator (ft³/in)	Air consumption per cycle per actuator (ft³/cycle)	Total air consumption for 8 cyclic operations (average) each of all actuators per machine [cfm (fad)]
Single-acting cylinder, 1¾", 2" stroke			
2	0.01002	0.02004	0.02004x8x2 = 0.32064
Double-acting cylinder, 2", 2" stroke			
3	0.0229178	0.0458356	0.0458356x8x3=1.1000544
Double-acting cylinder, 2½", 4" stroke			
2	0.0376506	0.1506024	0.1506024x8x2=2.4096384
Vacuum generator			
4	-	1.0	4.0
			7.8303328

Solution

Air consumption for 100% duty cycle per machine = 7.83 cfm

Actual air consumption for 10 machines = 78.3 cfm

Actual air consumption for 70% duty cycle = 54.81 cfm

Select compressor specifications from the manufacturer's chart (Table 3.3, Page 21)

- Compressor type = Reciprocating piston

- No. of stages = Single-stage

- Power, compressor = 15 hp

- Voltage = 440 V, three-phase

- Maximum pressure = 125 psi

- Tank size = 130 gallon

Design Problem 10.5

A manufacturing plant requires 10 Nos. of pneumatically-operated production machines connected to a ring-main distribution system. Each machine has to be designed for carrying out various work operations with the following specifications:

-Work operations, single-acting, 340 lb thrust force, 2" stroke, and 4 cycles per minute – 4 operations, and

-Work operations, double-acting, 560 lb thrust force, 2" stroke, and 6 cycles per minute – 8 operations

The operating pressure is to be set at 90 psi. The duty cycle can be taken as 70%. The compressor has to be located at a distance of 1000 ft from the ring main. The permissible pressure drop should be less than 14.5 psi. The necessary air quality level is 1.2.1 according to ISO 8573. Design a pneumatic system.

Solution		
Operating pressure (P)	90 psi	The pressure required at the point of use. Manufacturers rate the capacity of an actuator or tool at a specific pressure rating.
No. of machines	10	
For each machine:		
Single-acting cylinders		
No. of single-acting cylinders	4 Nos	
Force, single-acting cylinder (F)	340 lb	
Area of the piston ($A_P = F/P$)	3.78 in^2	[=340/90]
Bore diameter, [$D = \sqrt{(4 * A_P/\Pi)}$]	2.2 in	$= \sqrt{(4 * 3.78/\Pi}$
Bore diameter, (D), Selection (Table 7.1, Page 52)	2½ in	Std. bores (in): 5/16, 7/16, 9/16, ¾, 7/8, 1-1/16, 1¼, 1½, 1¾, 2, **2½**, 3
Rod Diameter, (d), Selection	5/8 in [0.625 in]	Bore dia (Rod dia) (in): 5/16(0.125), 7/16(0.187), 9/16(0.187), ¾(0.25), 7/8(0.25), 1-1/16(0.312), 1¼(0.437), 1½(0.437), 1¾(0.5), 2(0.625), **2½(0.625)**, 3(0.75)
Area of the piston, revised [$A_p = \pi D^2/4$]	$= \pi * (2.5)^2/4$ $= 4.9$ in^2	
Stroke length (S) (Given)	2"	
Air consumption per cycle per cylinder $\{(\pi D^2/4)*S*[(P+P_a)/P_a]\}$	$= (\pi 2.5^2/4)*2*[(90+14.5)/14.5]$ $=70.7535$ in^3 (std) $=70.7535/1728$ cu ft (std) $=0.0409$ cu ft (std)	(Also refer to Table 7.3, Page 55)
No. of cycles of operations per minute (Given)	4	
Air consumption per minute per cylinder	$= 0.0409 * 4$ $= 0.1637$ scfm	
Total air consumption of all single-acting cylinders (4 Nos) per machine	$= 0.1637 * 4$ $= 0.6548$ scfm	

Note: Atmospheric pressure is assumed as 14.5 psi

Double-acting cylinders		
No. of double-acting cylinders	8 Nos	
Force, double-acting cylinder (each)	560 lb	
Area of the piston (A = F/P)	6.22 in^2	
Bore diameter, [D = $\sqrt{(4 * A / \pi)}$]	2.81 in	
Bore diameter, (D), Selection	3 in	Std. bores (in): 5/16, 9/16, 1-1/16, 1½, 2, 2½, **3**, 3¼, 4, 5, 6, 8, 10, 12
Rod diameter, (d), Selection	5/8 in [0.625 in]	Bore dia (Rod dia) (in): 1-1/16(¼), 1½(⅝, 1), 2(⅝, 1), 2½(⅝, 1), **3(⅝)**, 3¼(1, 1⅜), 4(1, 1⅜), 5(1, 1⅜), 6(1⅜, 1¾), 8(1⅜, 1¾), 10(1¾, 2), 12(2, 2½)
Area of the piston, revised, (A$_p$) [A$_p$= $\pi D^2/4$]	= $\pi * 3^2/4$ = 7.065 in^2	
Effective area on the piston-rod side [A$_r$= $\pi(D^2-d^2)/4$]	= $\pi[3^2-(5/8)^2]/4$ = 6.7584 in^2	
Stroke length (Given)	2 in	
Air consumption per cycle per cylinder {($\pi D^2/4$) x S x [(P+P$_a$)/P$_a$]}+{[$\pi(D^2-d^2)/4$)] x S x [(P+P$_a$)/P$_a$]} =($\pi*3^2/4$) x 2 x [(90+14.5)/14.5] + [$\pi*(3^2-0.625^2)/4$] x 2 x [(90+14.5)/14.5]	=199.3479 in^3 (std) =199.3479/1728 cu ft (std) = 0.115363 cu ft (std)	D=3, d=5/8, S=2 P=90, Pa = 14.5
Air consumption per inch stroke of cylinder* [Please refer to Table 7.3, Page 55]	0.0576525 cu ft (std)	Reconcile
Air consumption per cycle per cylinder	= 0.0576525 x 2 = 0.115305 cu ft (std)	
No. of cycles of operations per minute	6	
Air consumption per minute per cylinder	= 0.115363 x 6 scfm = 0.692178 scfm	
Total air consumption of all double-acting cylinders per machine (x8)	= 0.692178 x 8 scfm = 5.537424 scfm	
Cushion	As required	
Stop tube	As required	
Dual pistons	As required	
Type	NFPA	ISO / NFPA
Bore section	Round	Square / Round
Cylinder mounting	Flange	[Fange, Foot, Pivot, Clevis, Trunnion]
Total Air Consumption for all machines		
Total air consumption of all actuators per machine	=0.6548 + 5.537424 =6.192224 scfm	
Total air consumption of all 10 machines	61.92224 scfm	
Duty cycle	70%	
Effective requirement of air	=61.92224 x 0.7 =43.345568 scfm	
Leakage (Assumed)	30% (~13 scfm)	
Total air consumption of all machines, considering leakage	=43.345568 x 1.3 =56.3492384 scfm	

*To estimate the total average air consumption of a pneumatic system, calculate the air consumption for each cylinder using the formulae given on Page 54. Values of air consumption for forward and return strokes of pneumatic cylinders are given in Table 7.3 on Page 55.

Compressor Unit & Drive Motor		
Operating pressure	90 psi	To be based on maximum pressure ratings of actuators
Maximum pressure drop (Given)	14.5 psi	
Working pressure, minimum	=90 + 14.5 psi =104.5 psi	
Working pressure, maximum	=104.5 + 14.5 psi =119 psi	
Actual Delivery, scfm(FAD)	56 scfm	Delivery: FAD / STD / Normal
No. of stages of the compressor	Single-stage	-Single compressor (for small and near constant compressed air demand) -Multiple compressors (for large and widely fluctuating demand, maintenance flexibility, backup compressor)
Compressor 1 (for single or multiple configurations)		
No. of stages of the compressor	Single-stage	
Type of compressor	Piston type	Piston/Screw/Vane
The temperature of the outlet air (Assumed)	250°F	Consult manufacturers' catalog
Compressor Drive supply	460 V, 3-phase	Consult manufacturers' catalog
Drive power (Selection), Table 3.3 Page 21	20 hp	Consult manufacturers' catalog
Drive Speed (Selection)	925 rpm	Consult manufacturers' catalog
Duty cycle	Continuous	
Drive, coupling	Belt	Belt/Gear/Direct
Lubrication	Lubricated	Lubricated/Dry run
Design	Open frame	Open-frame /Closed-frame
Noise level	70 dB(A)	As per the application requirement
Compressor 2	Not required	Applicable for multiple compressor system
Compressor 3	Not required	Applicable for multiple compressor system
Standby compressor	Not required	
Air receiver Capacity, Table 3.3, Page 21	120 gal	Consult manufacturers' catalog
Air receiver type (As per the application requirement)	Horizontal	Horizontal/Vertical
Receiver drain	Manual	
Receiver material	M S	Mild steel / Stainless steel
Certification for receiver tank	Relevant standard	
Pressure regulation	Motor start-stop control	Motor start-stop control / Variable frequency drive control
Pressure setting, the upper limit	104.5 psi	
Pressure setting, the lower limit	119 psi	
Compressor Control in compressed air station with multiple compressors	Not applicable	Cascade control / Pressure band control /Demand pressure control

Aftercooler

Aftercooler		
Aftercooler required	Yes	
If required, the type of configuration	Stand alone	Standalone /Package
Type of cooling medium	Air-cooled	Air-cooled / Water-cooled
Flow rate capacity@90 psi (at least)	56 scfm	
Ambient temperature (Assumed)	75°F	
Approach temperature (t_{amb}+15°F) Table 4.3, Page 32	75+15 = 90°F	t_{amb} + (5, 10, **15**, 20, or 30°F)
Outlet temperature, Compressor (See the previous page)	250°F	
Air inlet temperature rating, Aftercooler	≥250°F	
Fan rating (Selection)	0.5 hp	Consult manufacturers' catalog
Voltage	110VAC,1-ph	Consult manufacturers' catalog

Mainline Filter

Mainline Filter		
Filter Type	General-purpose	General purpose, Coalescing, or Adsorbent
Filter bowl	Transparent	Transparent (Polycarbonate) or Metal with liquid-level indicator
Flow rate capacity@90 psi (at least) of filter	56 scfm	
Quality Class to ISO 8573-1 (Customer requirement)	For particulates: 1	Given: 1.2.1
Filter element size	40 μ	
Service life indicator, filter	Mechanical	
Drain, for filter	Manual	
Pressure rating, filter	>120 psi	Mechanical or Electrical
Connection size, filter	¼	Manual or Automatic
Working pressure, filter	90 psi	
Connection size, filter	G¼	

Dryer

Dryer		
Dryer requirement	Required	Required / Not required
Quality Class to ISO 8573-1 (Customer requirement)	For moisture: 2 For oil: 1	Given: 1.2.1
Type of dryer	Desiccant (Unheated)	Desiccant (Unheated, Heated, Heated blower) / Refrigerant
Pressure dew point	-40°F [Class 2 for moisture]	Desiccant type: -94°F, -40°F, -4°F Refrigerated type: +37°F, +45°F, +50°F
Flow rate capacity@6 bar (at least)	56 scfm	
Working Pressure	>120 psi	
Inlet air temperature	95°F	
Heater capacity	NA	
Blower capacity	NA	
Connection size, dryer	G¼	
Pre-filter, particulate	1 μ	
Pre-filter, ultra-fine	0.01 μ	Coalescing filter
After -filter	Activated carbon filter	[Class 1 for oil]

FRL – Filter		
Filter Type	General-purpose	
Filter bowl	Transparent	
Flow rate capacity@90 psi of the filter	>5.6 scfm	
Filter element size (See Table 4.1, Page 28, Class 5)	5 μ	
Service life indicator, for filter	Mechanical	
Drain, for filter	Manual	
Pressure rating, filter (Table 4.4, Page 33)	>120 psi	
Connection size, filter	¼	
Thread type	NPT	

FRL – Regulator		
Regulator diaphragm type	Relieving type	
Requirement of gauge	With gauge	
Flow rate capacity of the regulator	>5.6 scfm	
Pressure rating, regulator (Table 6.1, Page 50)	120 psi	
Connection size	¼	
Thread type	NPT	

FRL – Lubricator		
Lubricator type	Micro-fog	
Lubricator bowl	Transparent	
Bowl capacity	0.5 liter	
Flow rate capacity of lubricator	>5.6 scfm	
Pressure rating (Table 6.2, Page 50)	>120 psi	
Connection size	¼	
Thread type	NPT	

Final Control Elements for Single-acting Cylinders		
Type (3/2-way Double solenoid valve/ 5/2-way double solenoid valve with port 2 blocked)	3/2-way Double solenoid valve	
Nominal flow rate, min	>5.6 scfm	
Valve connection size	¼	
Approximate Nominal Size (See Table 8.3, Page 65)	0.3 in	
Pressure rating (at least)	120 psi	
Seals	NBR	
Mounting Interface [DIN 24340 A6 / ISO 4401 / CETOP RP 121-H / NFPA D03]	NFPA	
Control voltage	24 V DC	

Final Control Elements for Double-acting Cylinders		
Type	5/2-way Double solenoid	
Nominal flow rate, min	>5.6 scfm	
Valve connection size	¼	
Approximate Nominal Size (See Table 8.3, Page 65)	0.3 in	
Pressure rating	>120 psi	
Seals	NBR	
Mounting Interface DIN 24340 A6 / ISO 4401 / CETOP RP 121-H / NFPA D03	NFPA	
Control voltage	24 V DC	

Conductor (Mainline)		
Total air consumption	56 scfm	
Length of pipe (Given)	1000 ft	
Pressure	120 psi	
Permissible pressure drop	15 psi	
Pipe diameter, the initial calculation (See Table 5.3, Page 46)	2 in	
No. of elbows	5 No	
No. of T-pieces	6 No	
No. of 2-way valves	1 No	
No. of corner valves	0	
No. of slide valves	0	
The equivalent length of one elbow (Table 5.4, Page 46)	5.7 ft	
The equivalent length of one T-piece	12 ft	
The equivalent length of one 2-way valve	10.6 ft	
The equivalent length of corner valves	0	
The equivalent length of slide valves	0	
The equivalent length of all elbows (x5)	28.5 ft	
The equivalent length of all T-pieces (x6)	72 ft	
The equivalent length of all 2-way valve (x1)	10.6 ft	
The equivalent length of all 2-way valve	0	
The equivalent length of all 2-way valve	0	
Total of equivalent lengths	111.1 ft	
Total of pipe length + equivalent lengths	1111 ft	
Pipe diameter, the final calculation	2 in	
Pipe diameter, a selection from the manufacturer's domain (Next same or larger size available)	2 in	

Figure 10.3, Page 83, gives the component-level layout of the system showing the essential design parameters.

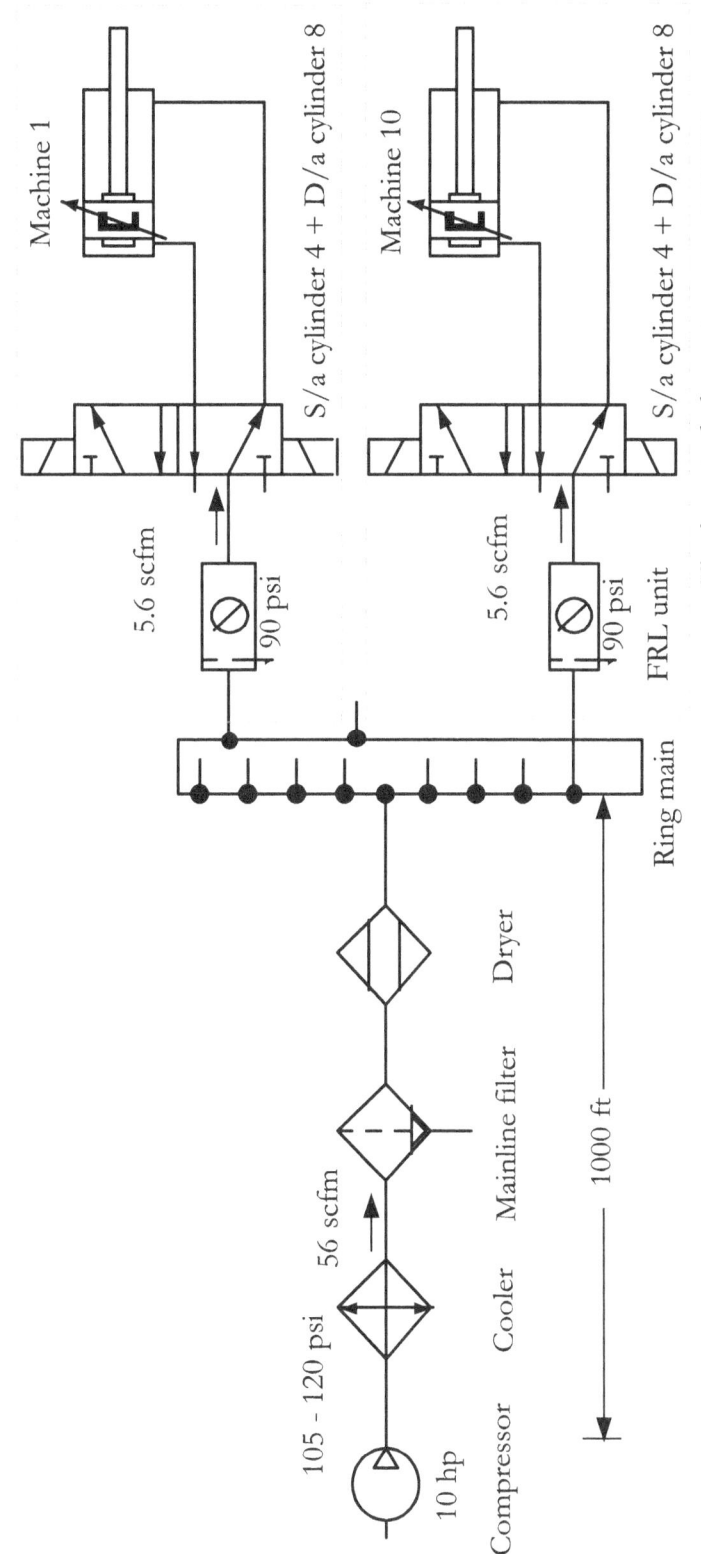

Figure 10.3 | A component-level layout of the system showing essential design parameters

Machine 1

S/a cylinder 4 + D/a cylinder 8

Machine 10

S/a cylinder 4 + D/a cylinder 8
actuator

Final control element

5.6 scfm

90 psi

5.6 scfm

90 psi

FRL unit

Ring main

Dryer

Mainline filter

105 - 120 psi   56 scfm

1000 ft

10 hp

Compressor   Cooler

Design Problem 10.6 | Demand-depended Compressed Air Delivery Splitting

A large-scale production system working in the early shift, day shift, and night shift demands compressed air at the following rates:

-Early shift   – 530 cfm (FAD) @110 psig
-Day shift   – 320 cfm (FAD) @110 psig
-Night shift   – 140 cfm (FAD) @110 psig

A standby compressor should be available as a backup unit for when a compressor needs to be taken offline for maintenance or replacement. Assume that the designer has the option to select rotary screw compressors with the following delivery rates:

1. 140 cfm (FAD) @110 psig
2. 160 cfm (FAD) @110 psig
3. 200 cfm (FAD) @110 psig
4. 280 cfm (FAD) @110 psig
5. 560 cfm (FAD) @110 psig
6. 46 to 220 cfm (FAD) @110 psig, variable speed drive

To realize the compressed air demand, draw various options for compressed air deliveries and compressor configurations, including a single compressor to multiple compressors, apart from the standby compressor. Also, suggest the probable efficiency of each option.

Solution

The maximum compressed air demand of 530 cfm (FAD) during the course of a day can be met with a single compressor or multiple compressors. Remember, a compressor operates efficiently when running at full load capacity. Various options can be worked out to configure the compressors with lesser or higher degrees of efficiency using multiple compressors. Some options are highlighted for study purposes.

Option 1

The maximum demand of 530 cfm can be met by using the rotary screw compressor with a delivery rate of 560 cfm. The compressor operates at its highest efficiency when the load is maximum, that is, during the early shift. The same compressor is used to meet the low loads in other shifts, and its efficiency tends to be low. The probable efficiency can be taken as 40%.

An additional rotary screw compressor with a delivery rate of 560 cfm can be used as a standby compressor.

Option 2

The maximum demand can also be met by using two rotary screw compressors with the following delivery rates:
1. 280 cfm (FAD)
2. 280 cfm (FAD)

-Two compressors with a total delivery of 560 cfm (FAD) must run during the early shift to meet the 530 cfm (FAD) demand.
-Two compressors with a total delivery of 560 cfm (FAD) must run during the early shift to meet the demand of 320 cfm (FAD).
-One compressor must run with a delivery of 280 cfm (FAD) during the night shift to meet the demand of 140 cfm (FAD).

The probable efficiency for this configuration of compressors can be taken as 60%.
An additional rotary screw compressor with a delivery rate of 280 lpm (std) can be used as a standby compressor.

Option 3
The maximum demand can also be met by using three rotary screw compressors with the following delivery rates:
1. 280 cfm (FAD)
2. 160 cfm (FAD)
3. 160 cfm (FAD)

-Three compressors must run with a total delivery of 600 cfm (FAD) during the early shift to meet the 530 cfm (FAD) demand.
-Two compressors must run with a total delivery of 320 cfm (FAD) [160 + 160 cfm (FAD)] during the day shift to meet the demand of 320 cfm (FAD).
-One compressor must run with a total delivery of 160 cfm (FAD) during the night shift to meet the 140 cfm (FAD) demand.

The probable efficiency for this configuration of compressors can be taken as 90%.

An additional rotary screw compressor with a delivery rate of 200 cfm (FAD) can be used as a standby compressor.

Option 4
The maximum demand can also be met by using three rotary screw compressors with the following delivery rates:
1. 200 cfm (FAD)
2. 140 cfm (FAD)
3. 46 to 220 cfm (FAD), variable speed drive

-Three compressors (including the variable displacement compressor) must run with a total maximum delivery of 560 cfm (FAD) during the early shift to meet the demand of 530 cfm (FAD). The delivery can be adjusted to meet the exact requirement using the VFD.
-Two compressors must run with a total maximum delivery of 360 cfm (FAD) [140 + ~220 cfm (FAD)] during the day shift to meet the demand of 320 cfm (FAD). The delivery can be adjusted to meet the exact requirement using the VFD.
-One compressor must run with a maximum delivery of ~220 cfm (FAD) during the night shift to meet the demand of 140 cfm (FAD). The delivery can be adjusted to meet the exact requirement using the VFD.

The probable efficiency for this configuration of compressors can be taken as 95%.

An additional rotary screw compressor with a delivery rate of 200 cfm (FAD) can be used as a standby compressor.

A summary of various configurations for the demand-depended delivery across compressors of different sizes is given in Figure 9.2.

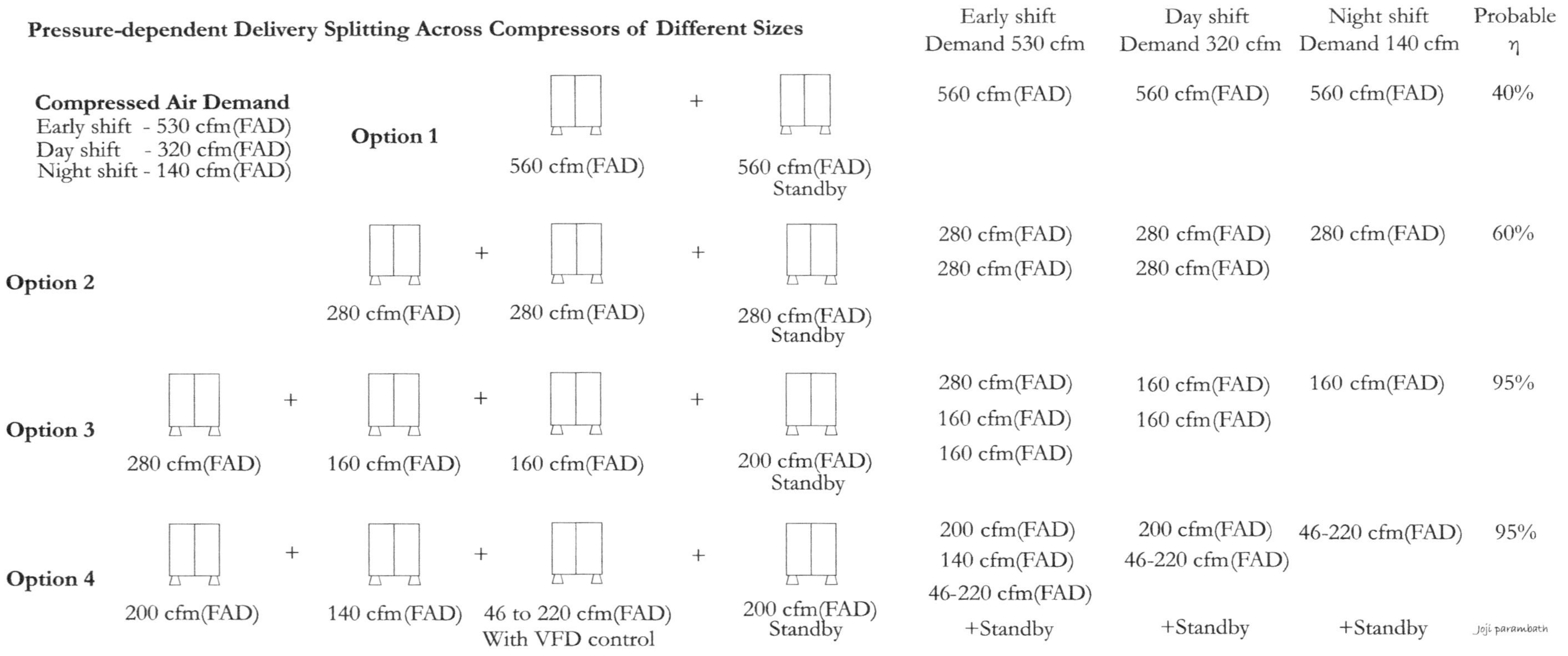

Pressure-dependent Delivery Splitting Across Compressors of Different Sizes

Compressed Air Demand
Early shift - 530 cfm(FAD)
Day shift - 320 cfm(FAD)
Night shift - 140 cfm(FAD)

	Early shift Demand 530 cfm	Day shift Demand 320 cfm	Night shift Demand 140 cfm	Probable η
Option 1	560 cfm(FAD)	560 cfm(FAD)	560 cfm(FAD)	40%
Option 2	280 cfm(FAD) 280 cfm(FAD)	280 cfm(FAD) 280 cfm(FAD)	280 cfm(FAD)	60%
Option 3	280 cfm(FAD) 160 cfm(FAD) 160 cfm(FAD)	160 cfm(FAD) 160 cfm(FAD)	160 cfm(FAD)	95%
Option 4	200 cfm(FAD) 140 cfm(FAD) 46-220 cfm(FAD) +Standby	200 cfm(FAD) 46-220 cfm(FAD) +Standby	46-220 cfm(FAD) +Standby	95%

Option 1: 280 cfm(FAD) + 560 cfm(FAD) + 560 cfm(FAD) Standby

Option 2: 280 cfm(FAD) + 280 cfm(FAD) + 280 cfm(FAD) Standby

Option 3: 160 cfm(FAD) + 160 cfm(FAD) + 200 cfm(FAD) Standby

Option 4: 200 cfm(FAD) + 140 cfm(FAD) + 46 to 220 cfm(FAD) With VFD control + 200 cfm(FAD) Standby

Figure 10.4 | Demand-dependent delivery across compressors of different sizes

Design Problems

1. An industrial unit requires 870 scfm of air at a continuous rate for its production process. The operating pressure is to be set at 100 psi. The compressor unit has to be located at a distance of 650 ft from the ring main. The permissible pressure drop should be less than 10 psi. The necessary air quality level is 3.4.2 according to ISO 8573-1 (2010). A standby compressor unit is to be provided to ensure continuous production. Design a suitable pneumatic system.

2. An industrial unit requires 195 scfm of air at a continuous rate for its production process. The operating pressure is to be set at 90 psi. The compressor unit has to be located at a distance of 1000 ft from the ring main. The permissible pressure drop should be less than 14.5 psi. The necessary air quality level is 2.4.4 according to ISO 8573. A standby compressor unit is to be provided to ensure continuous production. Design a pneumatic system optimally.

3. A large-scale production system working in the 1st shift, 2nd shift, and 3rd shift demands compressed air at the following rates:

   1st shift – 560 scfm
   2nd shift -420 scfm
   3rd shift – 280 scfm

   Draw two options for compressed air deliveries and compressor configurations, including a single compressor to multiple compressors, apart from the standby compressor, to realize the compressed air demand.

Appendix 1

Air Compressor Specifications

Specifications of Portable Electric Air Compressors

Table A1.1 | Specifications of portable compressors

hp	FAD cfm @90 psi	Tank size (gal)	Max pressure (psi)	Phase (Electric supply)	Input voltage (V, ac)	Remarks
0.33	0.6	2	100	1	115	--
0.3	0.75	1	135	1	115	--
0.5	0.85	2	125	1	115	--
1	2.35	2.3	125	1	115	--
0.75	2.6	3	135	1	115	--
0.8	2.6	6	150	1	115	--
0.9	2.6	6	150	1	115	--
1.1	3	2.5	200	1	115	--
2	3.3	2.6	130	1	115	--
1.1	3.4	4	125	1	115	--
2	3.8	4.2	125	1	115	--
1.5	4	1.5	125	1	115	--
1.1	4	4	125	1	115	--
2	4.1	4.3	135	1	115	--
2	4.2	3.2	125	1	115	--
2	4.6	4	200	1	115	--
1.3	4.6	4	125	1	115	--
1.8	4.9	6	125	1	115	--
1.8	4.9	20	135	1	120	H
1.8	4.9	26	150	1	120	V
2	4.9	30	135	1	115/230	H
2	4.9	60	135	1	115/230	V
1.6	5	4.5	200	1	115	--
1.6	5	15	200	1	120	H
1.6	5	15	200	1	120	V
2	5	20	200	1	120	V
1.7	5.1	15	225	1	120	V
2	5.1	24	125	1	120	V
2	5.5	15	135	1	120/240	H
2	5.5	20	135	1	120/240	H
2	5.5	20	135	1	120/240	V
2	5.5	30	135	1	120/240	V
1.8	5.5	9	135	1	120/240	H
1.5	6.3	7.8	125	1	120/240	H
2.5	6.5	5.3	130	1	115	
3	6.5	5.2	140	1	115	
1.5	7.3	20	135	1	120	H

1.5	7.3	9	135	1	120/240	H
3	9.1	30	135	3	200-230/460	H
3	10.2	20	135	1	240	H
3	10.2	60	135	1	230	V
3	10.3	60	135	1	230	V
3	12.7	9	150	1	240	H
5	14.2	60	140	3	208-230/460	V
5	14.2	60	140	1	230	V
5	15.5	60	135	1	230	V

H – Horizontal | V – Vertical

Specifications of Duplex Air Compressors

Table A1.2 | Specifications of duplex compressors

hp	FAD cfm @90 psi	Tank size (gal)	Max. pressure (psi)	Phase (Electric supply)	Input voltage (volt)	Remarks
5	32.8	120	175	1	208-230	H
5	32.8	120	175	3	208-230/ 460	H
7.5	49	120	175	1	208-230	H
7.5	49	120	175	3	208-230/ 460	H
10	72	200	175	3	208-230/ 460	H
15	100	200	175	3	208-230/460	H

H – Horizontal | V – Vertical

Specifications of Rotary Screw Type Single-stage Compressors

Table A1.3 | Specifications of rotary screw single-stage compressors

hp	FAD cfm @90 psi	Tank size (gal)	Max. pressure (psi)	Phase (Electric supply)	Input voltage (volt)	Remarks
5	18.2	80	150	1	230	H/V/E
5	18.2	80	150	3	208	H/V/E
5	18.2	80	150	3	230	H/V/E
5	18.2	80	150	3	460	H/V/E
5	18.8	9	150	1	240	H
7.5	27	80	150	1	230	H/V/E
7.5	27	80	150	3	208	H/V/E
7.5	27	80	150	3	230	H/V/E
7.5	27	80	150	3	460	H/V/E
10	36	80	150	3	208	H/V/E
10	36	80	150	3	230	H/V/E
10	36	80	150	3	460	H/V/E
15	46	80	150	3	208	H/V/E
15	46	80	150	3	230	H/V/E
15	46	80	150	3	460	H/V/E
20	68	120	150	3	208	H/V/E
20	68	120	150	3	230	H/V/E
20	68	120	150	3	460	H/V/E
25	87	120	150	3	208	H/V/E
25	87	120	150	3	230	H/V/E
25	87	120	150	3	460	H/V/E
30	117	120	125	3	208	H
30	117	120	125	3	230	H
30	117	120	125	3	460	H
40	184	Tankless	125	3	208	-
40	184	Tankless	125	3	230	-
40	184	Tankless	125	3	460	-
50	220	Tankless	125	3	208	-
50	220	Tankless	125	3	230	-
50	220	Tankless	125	3	460	-

H – Horizontal | V – Vertical | Enclosed

Specifications of 2-stage Air Compressors

Table A1.4 | Specifications of 2-stage compressors

hp	FAD cfm @90 psi	Tank size (gal)	Max. pressure (psi)	Phase (Electric supply)	Input voltage (volt)	Remarks
5	14	80	175	1	230	V
5	16.6	80	175	1	208-230	H
5	16.6	80	175	1	208-230	V
5	14	80	175	3	200	V
5	14	80	175	3	230	V
5	14	80	175	3	460	V
5	16.6	80	175	3	230/ 460	H
5	16.6	80	175	3	230/ 460	V
7.5	24.3	80	175	1	208-230	V
7.5	24	80	175	3	200	V
7.5	24	80	175	3	230	V
7.5	24	80	175	3	460	V
7.5	24.3	80	175	3	208-230/ 460	H
7.5	24.3	80	175	3	230/ 460	V
10	35	120	175	3	200	H
10	35	120	175	3	200	V
10	34.1	90	175	3	200-208	V
10	34.1	120	175	3	208-230	H
10	34.1	90	175	3	230/ 460	V
10	34.1	120	175	3	208-230/ 460	H
10	35	120	175	3	230	H
10	35	120	175	3	230	V
10	35	120	175	3	460	H
10	35	120	175	3	460	V
15	50	120	175	3	200	H
15	50	120	175	3	200-208	H
15	50	120	175	3	230/ 460	H
15	50	120	175	3	230/460	H
20	62	120	175	3	230/ 460	H
25	84	120	175	3	230/ 460	H
30	95	120	175	3	230/ 460	H

H – Horizontal | V – Vertical

Specifications of 2-stage Air Compressors

Table A1.5 | Specifications of rotary screw compressors

hp	Flow rate scfm @Operating pressure	Operating pressure (Maximum pressure) (psi)	Sound pressure level dB(A)	Input voltage (VAC)	H/V/E
25	112	110 (120)	65	3 ph, AC	E
25	93	145 (175)	65	3 ph, AC	E
30	138	110 (120)	66	3 ph, AC	E
30	110	145 (175)	66	3 ph, AC	E
30	91	190 (220)	66	3 ph, AC	E
35	162	110 (120)	66	3 ph, AC	E
35	136	145 (175)	66	3 ph, AC	E
35	108	190 (220)	66	3 ph, AC	E
40	195	110 (120)	69	3 ph, AC	E
40	159	145 (175)	69	3 ph, AC	E
40	131	190 (220)	69	3 ph, AC	E
175	892	110 (120)	74	3 ph, AC	E
175	720	145 (175)	74	3 ph, AC	E
175	570	190 (220)	74	3 ph, AC	E
215	1066	110 (120)	75	3 ph, AC	E
215	872	145 (175)	75	3 ph, AC	E
215	698	190 (220)	75	3 ph, AC	E
270	1337	110 (120)	75	3 ph, AC	E
270	1064	145 (175)	75	3 ph, AC	E
270	860	190 (220)	75	3 ph, AC	E
330	1490	110 (120)	76	3 ph, AC	E
330	1314	145 (175)	76	3 ph, AC	E
330	1047	190 (220)	76	3 ph, AC	E

Specifications of 2-stage Pressure-lubricated Air Compressors

Table A1.6 | Specifications of 2-stage pressure-lubricated compressors

hp	FAD cfm @90 psi	Tank size (gal)	Max. pressure (psi)	Phase (Electric supply)	Input voltage (volt)	Remarks
5	17.3	80	175	1	208-230	H
5	17.3	80	175	1	230	V
5	17.3	80	175	3	208-230/ 460	H
5	17.3	80	175	3	208-230/ 460	V
7.5	23.1	80	175	1	208-230	H
7.5	23.1	80	175	1	208-230	V
7.5	23.1	80	175	3	208-230/ 460	H
7.5	23.5	80	175	3	208-230/460	V
10	34.8	120	175	3	200-208	H
10	34.8	120	175	3	230/460	H
15	49	120	175	3	230/460	H

H – Horizontal | V – Vertical

Compressor Selector Charts

Table A1.7 | Compressor selector chart for pressures up to 100 psi

| Total air consumption, cfm | | Pressure setting, psi | | HP |
Average use	Continuous operation	Lower limit	Upper limit	
Up to 6.6	Up to 1.9	80	100	½
6.7 – 10.5	2 – 3	80	100	¾
10.6 – 13.6	3.1 – 3.9	80	100	1
13.7 – 20.3	4.0 – 5.8	80	100	1½
20.4 – 26.6	5.9 – 7.6	80	100	2
26.7 – 32.5	7.7 – 10.2	80	100	3
32.6 – 60.0	10.3 – 20.0	80	100	5
60.1 – 73.0	20.1 – 29.2	80	100	7½
73.1 – 100	29.3 – 40.0	80	100	10
100.1 – 150	40.1 – 60.0	80	100	15
150.1 – 200	60.1 – 80.0	80	100	20
201 – 250	80.1 – 100.0	80	100	25

Table A1.8 | Compressor selector Chart for pressures from 120 to 150 psi

| Total air consumption, cfm | | Pressure setting, psi | | HP |
Average use	Continuous operation	Lower limit	Upper limit	
Up to 3.8	Up to 1.1	120	150	½
3.9 – 7.3	1.2 – 2.1	120	150	¾
7.4 – 10.1	2.2 – 2.9	120	150	1
10.2 – 15.0	3.0 – 4.3	120	150	1½
15.1 – 25.9	4.4 – 7.4	120	150	2
26.0 – 39.2	7.5 – 11.2	120	150	3
39.3 – 51.9	11.3 – 17.3	120	150	5
52.0 – 67.5	17.4 – 27.0	120	150	7½
67.6 – 92.5	27.1 – 37.0	120	150	10
92.6 – 140	37.1 – 57.0	120	150	15
140.1 – 190	57.1 – 77.0	120	150	20
190.1 – 240	77.1 – 97.0	120	150	25

Table A1.9 | Compressor selector Chart for pressures from 145 to 175 psi

| Total air consumption, cfm | | Pressure setting, psi | | HP |
Average use	Continuous operation	Lower limit	Upper limit	
Up to 11.9	Up to 3.4	145	175	1
12.0 – 18.5	3.5 – 5.3	145	175	1½
18.6 – 24.2	5.4 – 5.9	145	175	2
24.3 – 36.4	7.0 – 10.4	145	175	3
36.5 – 51.0	10.5 – 17.0	145	175	5
51.1 – 66.0	17.1 – 26.4	145	175	7½
66.1 – 88.2	26.5 – 35.3	145	175	10
88.3 – 135.0	35.3 – 55.0	145	175	15
135.1 – 185.0	55.1 – 75.0	145	175	20
185.1 – 235.0	75.1 – 95.0	145	175	25

Appendix 2

General Information on Pneumatic Actuators

1. Essential Specifications of Actuators

The essential technical specifications for pneumatic actuators are given in Table A2.1.

Table A2.1 | Typical specifications for pneumatic components:

1.	Medium	Compressed air, filtered, lubricated
2.	Operation	Doubling-acting, air-cushioned
3.	Operating pressure	1.5 to 175 psi
4.	Operating temperature	-15°F to +175°C
5.	Linear actuators:	
5.1	Size	Piston rod diameters from 0.04 to 12 inches
5.2	Thrust	0.6 to 10000 lb (at 90 psi)
5.3	Stroke length	0.04 inches to 32 feet
5.4	Speed	0.2 to 600 in/s
6.	Rotary actuator:	
6.1	Size	Rotary drive diameter from 0.2 to 4 inches
6.2	Torque	1.3 to 1300 in-lb (at 90 psi)
6.3.	Angle of rotation	1° to 360°
6.4.	Speed	Up to 50000 rpm

Appendix 3

Air Consumption Chart for Industrial-type Tools

Table A3.1 | Air consumption at 70 to 90 psi

Tool	Consumption (cfm) at 25% usage factor
Air motor, 1 hp	9
Air motor, 2 hp	18
Air motor, 3 hp	24
Burring tool, large	6
Burring tool, small	4
Chipping hammer	8
Die grinder, medium	6
Drill, 1/16" to 3/8"	6
Drill, 3/8" to 5/8"	9
Horizontal grinder, 2"	8
Horizontal grinder, 4"	15
Horizontal grinder, 6"	18
Horizontal grinder, 8"	20
Impact Wrench, 1"	11
Impact Wrench, 1/2", 5/8", 3/4"	8
Impact Wrench, 1/4"	4
Impact Wrench, 1¼"	14
Impact Wrench, 3/8"	5
Nut setters, large up to 3/4"	15
Nut setters, small up to 3/8"	6
Paint spray gun	5
Rammers, medium /large	10
Rammers, small	6
Riveting hammer, Heavy	8
Riveting hammer, Light	4
Saws, circular	16
Scaling hammer	3
Screwdriver #2 to #6 screw	3
Screwdriver #5 to 5/16" screw	6
Trapper, up to 3/8"	6
Vertical grinders and sanders 5" pad	9
Vertical grinders and sanders 7" / 9" pad	20

Note:

- *Air consumption is only indicative and may not be accurate for any particular make*
- *Always check with the OEM for the actual air consumption of tools*

Air Consumption Chart for Automotive Service Shops

Table A3.2 | Air consumption of tools for automotive service shops

Tool	Consumption cfm (FAD)	Max. Pressure (psi)
Portable Tools		
Blow Gun	2.5	90
Body Polisher	20	90
Body Sander	10	90
Brake Tester	4	90
Burring Tool	5	100
Die Grinder	5	90
Drill, 1/16" to 3/8"	4	90
Filing and Sawing Machine	5	100
Impact Wrench 3/8" to 1" sq. dr.	3 - 12	90
Vertical Disc Sanders	20	100
TIRE TOOLS		
Air Hammer	4	100
Bead Breaker	12	150
Rim Stripper	6	150
Spring Oiler	4	100
Tire Hammer	12	100
Tire Inflation Line	2	150
Vacuum Cleaner	7	150
SPRAY GUNS		
Engine Cleaner	5	100
Paint Spray Gun (production)	8	100
Paint Spray Gun (touch up)	4	100
Paint Spray Gun (undercoat)	19	100
OTHER EQUIPMENT		
Car Lift, 8000 lb	6	175
Car Washer	9	100
Grease Gun	3	150
Medium Duty Sander	40	100
Spark Plug Cleaner	5	100
Transmission Flusher	3	100

Note:

- *Air consumption is only indicative and may not be accurate for any particular make*
- *Always check with the OEM for the actual air consumption of tools*

11 | References

1. Air Compressor Guide - Getting the Most for Your Money, How to Select and Protect Your Air Compressor Investment, Kaeser Compressors, Inc. P.O. Box 946, Fredericksburg, VA 22404, www.kaeser.com
2. Andrew Parr, Hydraulics & Pneumatics, A technician's and Engineer's Guide, 2nd Edition, Butterworth, Heinemann, 1998
3. Anthony Esposito, Fluid Power with Applications, 6th Edition, Prentice-Hall of India, 2006
4. Brochure 'NFPA Cylinders' NASON, 1307 S Highway 11 • Walhalla SC, www.nasonptc.com
5. Brochure 'Roundline Plus Stainless Steel Body Actuators 5/16" to 3" bore Single and Double acting actuators', IMI NORGREN, www.norgren.com
6. Compressed Air Engineering, Basic principles, tips and suggestions, KAESER KOMPRESSOREN SE, P.O. Box 2143 – 96410, Coburg, GERMANY, www.kaeser.com
7. Compressed Air System Guide, Designing Your Compressed Air System, How to Determine the System You Need, Kaeser Compressors, Inc. P.O. Box 946, Fredericksburg, VA 22404, www.kaeser.com
8. Compressed Air System Installation Guide, Layout Considerations for a Reliable, Energy Efficient, and Safe Compressed Air System, by Kaeser's Compressed Air and Engineering Experts, Kaeser Compressors, Inc.511 Sigma Drive, Fredericksburg, Virginia 22408 USA, www.us.kaeser.com
9. Compressed Air Treatment Guide - Meeting Your Compressed Air Treatment Needs, How to Select the Right Equipment for Your Application, Kaeser Compressors de Guatemala y Cia. Ltda., Calzada Atanasio Tzul 21-00, Zona 12, Complejo Empresarial, El Cortijo II, Bodega 501 01012 Guatemala
10. Document on 'Air and Water Cooled Aftercoolers', Ingersoll-Rand, Air Solutions, Heavy Industrial Systems, Strada Provinciale Cassanese, 108, I-20060 Vignate, Milano, Italy
11. Document on 'Desiccant Air Dryers - Heatless, Heated and Heated Blower', Ingersoll-Rand
12. Document on 'Heatless adsorption compressed air dryers, CD 110+, CD 150+, CD 185+, CD 250+, CD 300+ Basic control', Atlas Copco, www.atlascopco.com
13. Documents on (1) 'Rotary Screw Compressors, ESD Series, With the world-renowned SIGMA PROFILE, Flow rate 6.2 to 47.2 m³/min, Pressure 5.5 to 15 bar', (2) Rotary Screw Compressors, BSD Series, With the world-renowned SIGMA PROFILE, Flow rate 1.12 to 8.19 m³/min, Pressure 5.5 to 15 bar', (3) 'Rotary Screw Compressors, CSD(X) Series With the world-renowned SIGMA PROFILE, Flow rate 1.1 to 19.4 m³/min, Pressure 5.5 to 15 bar', (4) 'Rotary Screw Compressors, DSDX Series With the world-renowned SIGMA PROFILE, Flow rate 4.8 to 34.25 m³/min, Pressure 5.5 to 15 bar', (5) 'Rotary Screw Compressors, ASD Series With the world-renowned SIGMA PROFILE, Flow rate 0.89 to 6.39 m³/min, Pressure 5.5 to 15 bar', (6) 'Rotary Screw Compressors, DSD Series With the world-renowned SIGMA PROFILE, Flow rate 3.5 to 26.6 m³/min, Pressure 5.5 to 15 bar', KAESER KOMPRESSOREN SE, Coburg, GERMANY, www.kaeser.com
14. Dr E. h. Dipl.-lng. Kurt Stoll, Festo AG & Co., Esslingen, 'From the first applications of compressed air and its subsequent utilisation to the technical systems of today', 1997
15. Energy, SAVINGS in Compressed Air Systems, Kaeser Compressors, Inc. P.O. Box 946, Fredericksburg, VA 22404, www.kaeser.com
16. H Meixner & R Kobler, Maintenance of pneumatic equipment and systems, Festo Didactic, 1st Edition, 1977
17. H. Meixner & R. Kobler, Introduction to pneumatics, 2nd Edition, Festo Didactic, 1977
18. Hesse, Examples of pneumatic applications, Blue Digest for Automation, Festo, 1999
19. Hesse, Grippers and their applications, Blue Digest for Automation, Festo AG & Co, 1998
20. J P Hasebrink, R Kobler, Fundamentals of pneumatic control engineering, 3rd Edition Festo Didactic, 1989
21. Joji P., Pneumatic controls, Wiley India Pvt Ltd, New Delhi, 2008
22. Operating conditions and standards in pneumatics, FESTO
23. Pneumatics – 2000, Norgren product catalogue
24. SIMPLIFIED VALVE CIRCUIT GUIDE A GUIDE TO UNDERSTANDING PNEUMATIC DIRECTIONAL CONTROL VALVES, NORGREN
25. The standard ISO 8573-1, Compressed air – Part 1: Contaminants and purity classes
26. White paper 'Compressed air preparation in pneumatics' Festo AG & Co. KG, Adeline Konzelmann, Air Supply Product Management, Technical Product Support: Milorad Garic, www.festo.com

Fluid Power Educational Series Books

1. Pneumatic Systems and Circuits -Basic Level (In the SI Units)
2. Industrial Pneumatics -Basic Level (In the English Units)
3. Pneumatic Systems and Circuits -Advanced Level
4. Electro-Pneumatics and Automation
5. Design of Pneumatic Systems (In the SI Units)
6. Design Concepts in Pneumatic Systems (In the English Units)
7. Maintenance, Troubleshooting, and Safety in Pneumatic Systems
8. Industrial Hydraulic Systems and Circuits -Basic Level (In the SI Units)
9. Industrial Hydraulics -Basic Level (In the English Units)
10. Hydraulic Fluids
11. Hydraulic Filters: Construction, Installation Locations, and Specifications
12. Hydraulic Power Packs (In the SI Units)
13. Power Packs in Hydraulic Systems (In the English Units)
14. Hydraulic Cylinders (In the SI Units)
15. Hydraulic Linear Actuators (In the English Units)
16. Hydraulic Motors (In the SI Units)
17. Hydraulic Rotary Actuators (In the English Units)
18. Hydraulic Accumulators and Circuits (In the SI Units)
19. Accumulators in Hydraulic Systems (In the English Units)
20. Hydraulic Pipes, Tubes, and Hoses (In the SI Units)
21. Pipes, Tubes, and Hoses in Hydraulic Systems (In the English Units)
22. Design of Industrial Hydraulic Systems (In the SI Units)
23. Design Concepts in Industrial Hydraulic Systems (In the English Units)
24. Maintenance, Troubleshooting, and Safety in Hydraulic Systems
25. Hydrostatic Transmissions (HSTs) (In the SI Units)
26. Concepts of Hydrostatic Transmissions (In the English Units)
27. Load Sensing Hydraulic Systems (In the SI Units)
28. Concepts of Load Sensing Hydraulic Systems (In the English Units)
29. Electro-hydraulic Proportional Valves
30. Electro-hydraulic Servo Valves
31. Cartridge Valves
32. Electro-hydraulic Systems and Relay Circuits
33. Practical Book: Pneumatics - Basic Level
34. Practical Book: Electro-pneumatics - Basic Level
35. Practical Book: Industrial Hydraulics – Basic Level
36. Programmable Logic Controllers and Programming Concepts
37. Compressed Air Dryers
38. Hydraulic Circuits – Identification of Components and Analysis

For more details, please visit: **https://jojibooks.com**